# The Secret Oceans

# The Secret Oceans

## Betty Ballantine

### Featuring the art of

| | |
|---|---|
| Lloyd Birmingham | Gilles Malkine |
| Steve Brennan | Jeffrey Mangiat |
| Joseph DeVito | Thomas McNeely |
| David Henderson | Davis Meltzer |
| Carol Inouye | Charles Passarelli |
| Robert Larkin | Jeffrey Terreson |

DORLING KINDERSLEY

Creative Consultants:
    Ian Ballantine
    Richard Ballantine

Book & Jacket Designer:
    Fearn Cutler de Vicq de Cumptich

Assistant Art Director:
    Gilles Malkine

The original spelling and punctuation of this story have been retained in an effort to achieve a unified language between humans and other species.

First published in Great Britain in 1994
by Dorling Kindersley Limited,
9 Henrietta Street, London WC2E 8PS

First published in the United States by Bantam Books

A CIP catalogue record for this book is available from the British Library.

ISBN 0-7513-5253-5

Printed in Italy by Amilcare Pizzi, s.p.a.

For my Grandchildren

I would  like to acknowledge
with affectionate gratitude the contribution of Ian Ballantine,
whose idea was the genesis for this book, and Richard Ballantine,
whose enthusiastic expertise filled many gaps; and special thanks
to Gilles Malkine and Fearn Cutler, whose steady support through all
vicissitudes saw the book to its completion.

# THE TURTLE IS READY!

The *Turtle,* a unique deep-sea submersible capable of extended dives at great depths, has completed sea trials and crew training and is now ready to undertake the most extraordinary project in the history of oceanic research—to communicate with the whales.

The wide array of advanced equipment on board the *Turtle* includes a state-of-the-future computer and audio-visual translator with which the expedition hopes to achieve *two-way communication with the whales.* The computer will interpret the thousands of organized sounds these behemoths make into sequences that make sense in human terms. The computer will then, through the talents of the two very special young aquanauts, Donna Christie and Paul Dufours, *interpret human sounds back to the whales.*

For the first time, mankind will actually "speak" to a different species, not just in another language, but in a whole new sound form.

No expense has been spared in setting up this expedition. The experimental equipment includes the *Shuttle,* a two-person, self-powered submersible designed to go deeper than any vessel has gone before. It can operate independently, or in conjunction with a robot probe submersible (affectionately known as the *Bug*), controlled from the mother ship. Each member of the crew assembled for this epochal experiment is a scientist, most in several disciplines.

**"The Bug"**

The captain, Katerina Kominsky, is retired from the disbanded Soviet Navy where she commanded a nuclear submarine. Kate, as she is known to everyone, has degrees in geophysics, computer technology and engineering, plus command of four languages.

Kate's assistant, Chief Engineer Teiji McTavish, comes festooned with degrees from universities as far apart as Japan and Scotland, and several in between. Languages: English, Japanese, Tagalog.

In command of all scientific aspects is, of course,

Captain Katerina Kominsky....

David Schlessinger, Chief of all scientific experiments

Paul Dufours and Donna Christie. Children of many talents....

doctor-scientist-professor David Schlessinger, world-famous for his head-on encounters with the despoilers of our planet, for his several dozen outspoken books on all manner of subjects, and for the quirky personality that has endeared him to millions of armchair adventurers who share his views.

It was David's insistence that ensured the participation of the expedition's two youngest members, Paul Dufours and Donna Christie, his genius-level adopted children. Paul's musical talents, at age fourteen, easily rival Mozart's and his

eight languages include Tibetan.

Donna Christie, an electronics whiz, at fifteen, has contributed significantly to the development of the *Turtle's* computer.

The combined talents of these two young people, music and computer technology, is crucial to the expedition's basic objective—to communicate with one of the planet's last great mammals.

The sixth member of the crew is our own correspondent, Jay Hunt, with a wide range of science writing for many disciplines acquired through hands-on experience. Jay is a reporter who believes in doing what he is writing about. His official duty on board is photography, combined with emergency medical skills gained in war combat.

While each member of the crew has his or her own assortment of specialties, all are capable of running the ship, and all will stand watch-on-watch once they are under way.

The *Turtle's* exact location is a very well-guarded secret. Professor Schlessinger was colorfully clear about not wanting any tourist ships in the area to create noise.

Tomorrow is the BIG DAY, the day when a high-tech modern miracle, designed to put humankind in communication with another species, will set out on its daring odyssey. Bon voyage!

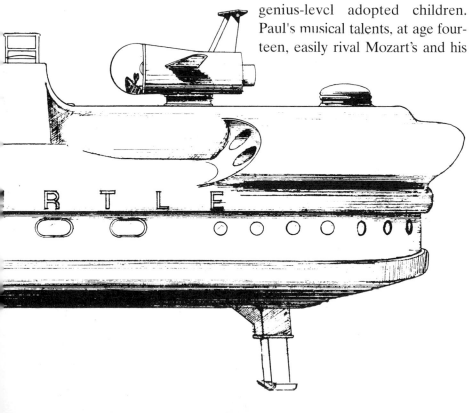

"The Shuttle"

On board the *Turtle*, unaware that they were being keenly watched, six people went their various ways about distracting themselves from pre-dive tension. Kate, satisfied that all systems were in order, settled in to a relaxing game of after-dinner chess with the Professor. Dave Schlessinger, passionate enthusiast in whatever he did, had no fears or qualms for the morrow, but was keyed up with anticipatory excitement. However, right now his attention was strictly on the game. He was always able to focus his wide-ranging mind to one fierce concentration whenever he wanted.

Teiji McTavish, on the other hand, normally cheerful, volatile, and always brilliant at his job, was over-tired, worn out by testing systems all day, even though the perfect precision of his beloved machines was an unending joy. He was buried in the old-fashioned navigational maps he found so soothing when he heard the strains of a flute, shortly joined by a sweet counterpoint soprano. Teiji cocked his head, listening. Then he grinned. Paul was definitely getting a syncopated beat into that Bach, Donna effortlessly following. Quickly the Chief Engineer bundled his maps, checked the autopilot, and made his way aft to join what had become a trio, for Jay Hunt was providing a soft, thrumming background on his mandolin. Now Teiji chinned his fiddle and joined in, while Donna abandoned singing for her Irish harp.

Most of the instruments aboard the *Turtle* were chosen to match the sounds of whales. And of course Paul could produce any sound he wanted on his synthesizer.

All six aquanauts were musically sophisticated, if only as hobbyists. Quite aside from the mutual trust and respect generated by shared expertise in running the diveship, the music was a personal expression that had helped to pull them together during the preceding weeks of living in close quarters. Somewhat to their own surprise, they had come to function more or less as a family. Even Jay and Teiji, both loners, found themselves relying on the always stable comfort that Kate offered, and responding to the buoyant sense of adventure in David. And all of them tended to worry needlessly over Paul and Donna, whose mercurial extremes were only emphasized by the superlative talents that had always set them apart.

Presently the group was joined by Kate and Dave, he wearing a distinct look of triumph. "I won!" he announced over the sounds of the music. "I won!" And seizing Kate, he whirled her round and round in the restricted space. Paul brought the music to a rapid close.

"Wow!" Dave was enthusiastically breathless, Kate laughing.

"Now what was that?" Dave demanded.

"Bach," Paul responded straight-faced.

"Oh, of course, Bach," Dave at once agreed. "I love that classical stuff. Greatest dance music ever written." It was a standard joke that while Dave had a keen ear and would harmonize at the drop of a note, he pretended to total ignorance about music. Now he launched them into a roaring series of sea songs—ribald, jovial and joyously raucous. Paul, well aware that his step-parent could easily sing the night away, skillfully steered the group down from their high to the wistfully plaintive melody of a woman's plea for her man to come home from sea.

As the song ended, Kate rose. "Okay, everyone, I think it's time to stop. Tomorrow's going to be a long day." She patted Paul. "Thank you, Paul," she murmured, and he knew she had understood his musical manipulation. The group dissolved among affectionate goodnights.

No one had any idea of just how long the next day would be.

The next day, with the dive well under way, Kate sat forward in her command chair, still enthralled at the miracle unfolding before her practically within touch. Learning how to scuba had been a joyous eye-opener for her. Nothing in her submarine experience had ever given her a hands-on, ringside seat at the undersea world.

For this first deep dive she had taken them down to 5,000 feet before ordering the ship to hover, all exterior lights out, so that they could listen. In the dim glow from their consoles, the crew looked out into a black world lit only by the wink and flash of darting bioluminescent forms.

The myriad sounds of the ocean, carried five times louder and faster by water than by airborne sound, invaded the stillness of the command module. They were assaulted by clicks, rattles, whispers, yelps, whistles, tappings—then a loud scraping.

Involuntarily Kate glanced at the monitor screen, but of course it was blank: with no outside lights, the cameras were not functioning. Everyone was motionless, listening, Paul with earphones on. They all heard the sound when it came—a soft, plaintive, piercingly long-drawn-out call, followed by a deep grunt.

"It's them!" An excited whisper from Paul. "It's the whales, they've found us!"

"Recording," Donna's quiet voice.

Now the call had sunk to a brassy rumble. This close, the deep tones took over the whole ship, reverberating up through the soles of their feet. They were enveloped by the sound, the winking lights in the black water taking on the same cradling rhythm. Kate was mesmerized, swaying with the sonorous cadence. Paul took off his earphones to pick up his flute, but then let it hang slack, abandoning himself to the subtle weaving of the whalesong.

Kate shook herself. "Teiji, let's have the lower port light, please." The world beyond the bubble leapt into sight. Clear in the beam they could see part of a humpback and hundreds of darting fish. Then there was a sudden lurch, and all the hurrying fish were blotted out.

Kate heard a small squeak from Donna before the light also went out.

The captain swore roundly in Russian.

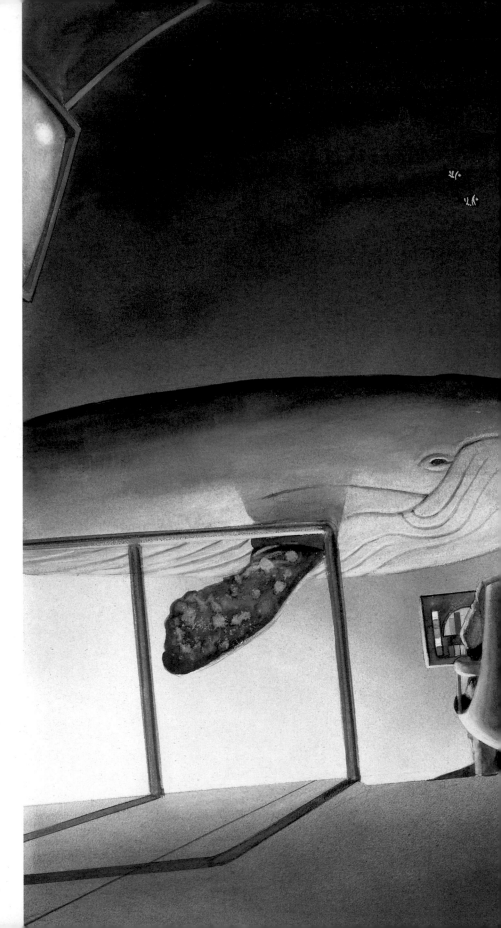

Dave's voice came quietly from the gloom behind her. "That's the mother of all giant squid, Kate. You aren't going to get rid of it easily."

"I don't give a damn what it is. I want it off my ship! Teiji? Lights!"

"Coming on line right now, Captain."

The squid slid away from the light, clearly disturbed even by the diminished glare of the auxiliaries.

"What's happened to the power?" Kate demanded.

"They must have blanketed the thrusters, Captain," from Jay.

Kate turned sharply. "*They?*"

"There has to be more than one," David said thoughtfully. "This is really most remarkable. What's more, I could swear we're moving."

They could all feel it—a slight tremor in the deck, and the flickering lights in the water, once more visible, were beginning to streak. Kate suddenly realized the whalesong was continuing. She muttered to herself, and then to Teiji, "Can you give me our speed?"

"No, Ma'am. We appear to be in a current, and if we are moving at the same speed as it, our effective speed is nil."

"And we could slam into something just as effectively hard," she commented.

It was a submariner's nightmare—no control and running blind. For Kate, the worst scenario had always been loss of command. By way of compensation, when circumstances took things out of her hands, Kate's own center of self control settled into total calm. It was one of the qualities that made her a good captain.

"Donna? Are you getting anything?" Her voice was almost casual.

"Well, sonar indicates random objects which we are avoiding."

Dave grunted. "That has to be deliberate. And ridiculous. Giant squid do not get together to capture and steer anything, much less an inedible ship."

In the near darkness (the *Turtle's* emergencies threw only restricted light) the nightmare journey continued. As did the whale song.

"As long as the whales are with us, we're not likely to descend much further," Paul remarked, surprising Kate with the calmness of his tone. She glanced around her command module. In fact, each member of the neophyte crew seemed amazingly at ease. Only the tense set of Teiji's shoulders, as he sat helpless in his pilot's seat, indicated his frustration.

"Actually we've been on a shallow upward tilt for some time now," he commented, confirming Kate's assessment.

Just then Donna exclaimed, "Kate, I'm getting something. I have sonar readings from all around."

"Distance?"

"We have a clearance of approximately 12 meters, varying to 17."

"Sounds like a tunnel," Dave remarked. "Could be a gigantic lava tube. Something seems to know what it's doing—" he broke off. "By all the gods! Kate, everyone—d'you realize what that means?"

"Intelligence," Paul said at once.

"Of course, intelligence!" David was vehement. "But whose? If only we had some light to see what the hell's going on. Squid are pretty

intelligent but there's no way they could have organized all this. Something has to be directing them—"

"David. You're letting your imagination run away with you," Kate's voice was quiet but firm. "However, I agree about the lights. Teiji? Anything?"

The Chief Engineer replied with obvious relief. "Yes, Captain. Thrusters show gain, we should be getting power momentarily."

And indeed, the main lights flickered and at last came on full, to reveal an absolutely amazing sight.

A concerted gasp greeted the startling appearance of three dolphins in the sudden light, two full grown and one small.

"Wait a minute," David said. "That's very odd. They're very weird dolphins—d'you see the crests on their heads?"

"Yes, like a pretty frill," Donna said, pleased. One of the creatures promptly waved.

"Hey!" from Jay. "That was a wave! Look, it really is waving at us!" And exultantly, "I've got it on video!"

The second dolphin-like animal, the larger one, had a spiny, fan-shaped crest, while the little one had tiny nobs and a small, white fleck by its eye.

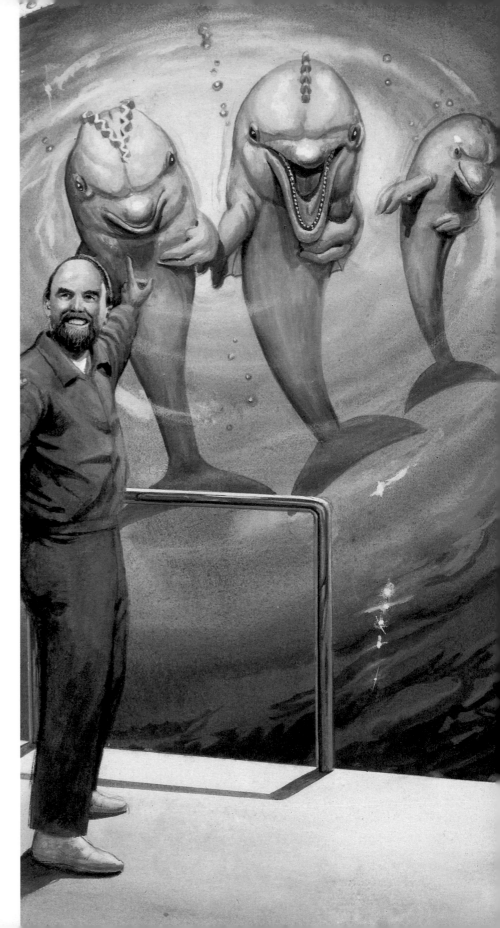

Donna was enchanted. "Oh, he's cute!" she exclaimed, not even seeing Paul's look of disgust at her reaction. But David agreed, and added a comment, "Donna, what's significant is that they feel safe enough to bring their young with them, despite those giant squid."

"I think they left just before the main lights came on," Teiji murmured. They were still moving very slowly, but no one gave a thought to how. Except Kate, and she kept quiet. The attention of the entire crew was riveted on the three creatures weaving ahead of them. Now the three drew up, back-watered, and bowed in unison. Teiji half-rose and bowed back. Glancing over his shoulder he hissed, "Bow back! Bow back!"

"Of course," David responded immediately. "Imitative body language. They're clearly trying to communicate. Kate, this is absolutely wonderful!"

"I'm glad you think so," Kate's voice was neutral. "We appear to be surfacing." And in the next few moments, the top of the *Turtle's* bubble emerged to reveal a huge lagoon, as big as a football field, lit by shafts of sunlight, and with a number of caverns leading off from it.

No one said anything for a time, awed by the underground landscape, and intensely relieved. But the newfound "dolphins" demanded attention. With body language they quickly established *headnod* for "yes" and *headshake* for "no". Dave was in a transport of excitement, officially dubbing the new species *cetasapien*. There was a corresponding flurry of excitement in the pool when Paul imitated the whalesong on his synthesizer. Then he tried variation after variation. Donna shook her head. "Keep it simple," she said. She had already coded in "CT" for cetasapien.

"But the song *isn't* simple," Paul argued. Doggedly they went at it hour after hour. There was not much the others could do, frustration eating them to exhaustion. Eventually Kate called a halt, decreeing a meal, sleep (the late hour surprised everyone), and a new start the next day.

Tired as she was, Donna found herself unable to sleep. She was not accustomed to failure. Ever since she had set out to attract the attention of a sympathetic teacher who got her out of the orphanage and into a school for gifted children, Donna had relied on herself, and, as an extension of herself, her computer, to solve problems.

David Schlessinger, a sponsor of the school, quickly recognized the quality of intelligence in the independent little girl, and added her to his adopted family, as a slightly older companion for the small boy whose extraordinary sensitivity and musical talent had created some difficulties. Donna had immediately become fiercely protective of Paul, "mothering" him in a bossy sort of way, completely unaware that of late he regarded her authority with some amusement.

Now she heaved a sigh, brooding over what she might try with her computer the next day, when she was startled to see Frill, as she had named the smaller of the two adult cetasapiens, appear at her window.

Donna immediately put her hand on the heavy, specially developed reinforced acrylic that allowed vision without distortion. Frill matched her gesture and the girl noticed the difference in the cetasapien "hand," tracing its three-fingered conformation with her own finger. Frill pointed at her, and waved toward the command module, pointed and waved repeatedly.

Donna was fascinated, exhaustion forgotten. Kate being fast asleep, characteristically the girl acted on her own, climbed out of bed, got dressed, and quietly made her way through the saloon to her console.

Frill was waiting for her at the bubble. Again she pointed, clearly indicating the computer. Donna sat down, one eye on the cetasapien: she tried various combinations, responding to nods or headshakes from the alert watcher. Nothing but garble appeared on her screen while the translator produced only deep rumblings and squeaks. Finally Donna signaled "wait," slipped aft, and woke a sleepy Paul.

"Come on," she whispered urgently, "I've got Frill working with me. We need sound—I think. The computer seems to have digested a whole lot while we were resting."

Paul came to full alert, and followed Donna in a few minutes, discovering that she and Frill had been joined by a whale.

He had become adept at reproducing the sounds of both whales and cetasapiens. Together they worked through different sets from the synthesizer in combination with computer recordings, echoing sound. One of the problems, he knew, was that the sea mammal sounds covered a much wider range than the human ear could compass.

Then Frill apparently realized they had to reduce their range to something like cetasapien "baby talk." At last, with the help of the computer, they achieved a median "A" which appeared on the screen as a particular resonance. From there Paul was able to project a series of sound symbols that translated as letters of the human alphabet. This way, the computer could project letters typed in as sound frequencies on the monitor, and in turn, interpret the cetasapien sounds as letters.

And suddenly they had it, there on the screen in front of them.

Donna gave a whoop, hugged Paul and rushed over to where Frill pressed hard up against the bubble. Paul practically stood on the button that roused the whole ship and brought the crew streaming into the command module.

It took a while for things to calm down enough for explanations to be made. David was avid to know how the contact had occurred, hanging over Donna's shoulder while the laborious process of spelling out messages took place. It was getting light enough for them to see that the water was fairly boiling with excited cetasapiens.

Paul threw up his hands. ONE, he played, ONE AT A TIME. Promptly a message came back, JOIN US. "Of course!" David at once endorsed. "We must get into the water with them right away. Jay . . ."

"Now, wait a minute, David," Kate was definitely alarmed. "You can't just leap in like that. You don't know what the hell is out there besides your CTs. Nor do you even know for sure what they intend."

"That's absurd. They've made it clear they are not hostile."

"How about being captured," Kate snapped. "You call that friendly? And what about those damn squid? You want to cuddle up to one of them?"

Dave was exasperated. "Kate, what's come over you? Surely I don't have to point out that the whole reason we're here is to communicate?"

"We can do that perfectly well from the safety of the ship!"

"Not really," very quietly from Paul. "This is a clumsy method: we need to work out something better. *With them*. Besides, our safety really depends on them anyway. They brought us here and we'll probably need them to get us out."

"Exactly. We've been trapped. That's just my point. Teiji, get up to the conning-tower and see what's out there."

"Katy," Dave was now being very patient, "you're letting your personal sense of responsibility for us distort your judgment . . ."

"Oh, look!" Donna interrupted. She had ignored the altercation, apparently intent on the cetasapiens. But she was perfectly aware of what was happening, and determined to support David. Now she innocently continued, "We've frightened them away."

Actually Frill, Fan and the small Fleck were still there, joined by an even smaller specimen who flicked back and forth, frolicking in the water right beside the captain's station.

"Kate," David was quietly urgent, "we have simply got to find out about this. Even at the cost of some risk."

"Right," Jay spoke up, "that little critter is making a play for you, Kate. Couldn't be clearer." Teiji reported over the intercom, "Captain, all clear from up here," and Jay continued smoothly, "so why don't I go in?"

"All right," Kate finally conceded, "but in full gear, including net knives. There's no telling what you'll encounter out there."

Jay rapidly went aft, and assisted by Paul, suited up and made his way into the lock chamber. From there he cautiously entered the water.

Actually, when Donna spoke of having frightened the cetasapiens away, she was closer to the truth than she knew. They could sense whole layers of emotions; deceit of any kind simply did not exist for them. Moreover, they tended to shy away from anger. And since their presence seemed to agitate the humans, most had simply disappeared.

Meantime Donna, never having really had any doubts about the cetasapiens, had been carefully watching the signals Frill was making. She decided Frill wanted her to come into the water too, and while the attention of the group was centered on what was happening with Jay, Donna slipped back to her cabin and changed into a swimsuit. Nipping up the ladder topside, she discovered Teiji had already opened the upper hatch. Running past an astonished Chief Engineer, she skipped to the edge of the *Turtle* and unhesitatingly dove in.

Emerging from the *Turtle*, Jay looked carefully about. The water was quiet, clear—and empty, lit by long shafts of light from the rising sun coming in through collapsed areas of the ancient caldera far above, and penetrating the lagoon at a sharp angle. Slowly he swam clear of the *Turtle* and made his way forward underwater.

The instant he got near the bubble, the smallest cetasapien flicked over to him and virtually hurled itself into his arms. The adults held back, watching, waiting for him to come to them. Wryly Jay recalled writing an article for an ethologist who, in contacting wild creatures, always waited to let them come to him. Jay felt complimented: the CTs respected him as wild, but trusted him enough to allow the small, affectionate young one frisking around him to come close. Gently, he approached the adults.

With great relief, Kate was watching Jay make friends with the cetasapien family. He had just swum over to some rocks to remove his mask and shuck his tanks when Donna appeared, swimming joyous circles with Frill.

Kate gasped and turned to David. Laughing out loud, he patted her shoulder and urged her toward the saloon.

"Come on, Katrina!" David was jubilant. "Let's get in there with them!"

Shortly thereafter every member of the crew was swimming, gratefully relaxed in the welcoming water of the lagoon, each one soon joined by an equally welcoming group of cetasapiens.

After a couple of hours of celebrating in the lagoon, not only with their newly found friends but, to everyone's amazement, with several other varieties of sea creatures, Teiji and Paul, somewhat reluctantly, returned to the *Turtle* to work on a better method of communicating with the CTs. Paul was determined to be able to talk directly, if only in a limited way, and more importantly, to do so while in the water.

He and Teiji worked through endless combinations on their digital wrist coms, which could both receive and transmit.

Eventually they were interrupted by the return of the rest of the crew, accompanied by CTs bearing fresh caught fish, concerned for the welfare of their happy "captives."

During lunch, every mouthful of which was eagerly watched by a large and very curious group of cetasapiens, David explained that if the CT sensory organs were like those of the upper air dolphins, they could probably see the food going down everyone's gullets.

"How come?" Jay wanted to know.

"Dolphins, and by extension CTs, don't see the way we do. In fact, 'see' is misleading. Human perception is very limited by comparison with what CTs can sense with echo-location, sonar, taste, hearing, and who knows what other biological senses."

He broke off, observing a lot of excitement out among their audience. Donna, seated by the window, glanced back at him and noticed the messroom screen.

"Oh, look!" she exclaimed, "Look what's happening!" On the screen shadowy images were moving, like photos printed one on top of another.

Jay jumped up. "Hey, something weird is going on with my cameras—they're printing multiple images. I'd better check—"

"No, wait. Maybe," David paused, slid a glance at the window, "could it be that they're projecting images? Maybe trying to show us how they 'see,' in 3-D as it were, including density, form, background—a whole range of sources that create a hologram, not just a flat surface?"

"But that's marvelous," Teiji said, "that could be a huge help in talking with them. Think of it—in-depth visuals instead of words."

"Not 'instead of.' We could have sound too," Paul commented, then falsetto'd, '*If I hadda talking picksher—of you-hoo.*'

"Shut-up," Jay said mildly. "Dave, you said they were projecting. How?"

"I said they might be. But I can't tell you how. It could be those crests: those may be some kind of evolved acoustic antenna. They must have an ability to broadcast, other than through sound."

Once again, the CTs were churning in high excitement. Then all but Frill flipped away. She came right up against the window, her crest rippling and weaving, folding flat to her head, only to reappear, quivering. And on the screen too, she was up close, but there the empty waters behind Frill were alive with members of her family.

"Lord," Kate breathed, "if they always communicate in pictures they can convey acres of information—zip!—just like that. No wonder they had trouble with simple letters."

"That's not all," David murmured. "Why didn't they do this before? And Katy, how did Frill know what I was talking about? After all, I don't have a crest."

Donna looked at him, wide-eyed. "She heard you, of course. You said yourself she can see right into your head." Abruptly Donna yawned. Kate took instant charge. "Donna, you and Paul need sleep. You were both up all night. Let's tackle this again tomorrow." Outside, Frill nodded.

Teiji and Jay spent the afternoon continuing to adapt the wrist coms, until, within a limited range, they were effective two-way devices. They'd work better underwater, of course, but even in open air, talking with CTs was going to be much easier.

The next day Frill led them to a large inner cave, evidently a common meeting place. Several of the many cetasapiens present crowded around the humans.

"My family," Frill announced proudly.

It became obvious that the cetasapiens were both air and water breathers: they had already demonstrated their ability to remain underwater for long periods of time. Now they all appeared to be thoroughly comfortable in the warm, moist air of the cavern. Everyone proceeded to get acquainted, Donna delightedly assigning names. They already knew Frill, her mate Fan, and the two young, Fleck and Flick. Now they met Uncle Tatty, Aunt Em, and several others. Paul and Donna went off with Frill's relations. But David was getting very impatient, and at last launched a barrage of questions.

Frill gave a throaty sound that translated as a chuckle. "Well, first," she said, "our crests are exactly as you surmised—an extra acoustic membrane that allows us to sense and to send. Especially in water." She paused. "We notice you address each other by name-sounds: your young girl has already assigned such recognition symbols to several of us. Between ourselves, we do not need such symbols. Each of us is instantly recognizable to all by the unique image we convey."

"But who are you? What are you? Why are we here?"

Frill shook her head. "Patience. You will learn about us, yes. And we must learn about you also. There is much to see. Questions will take care of themselves," cocking her head at them, "from looking. In all ways."

Kate glanced at David. "You know what? I think they exchange information in such wholesale form that simple questions just puzzle them."

Frill nodded vigorously: "Your female leader Kate is right. Tomorrow the home cave," she announced.

Getting to Frill's home lagoon involved a fairly long swim in dive gear through what the humans came to learn was a labyrinth of caves, the underpinnings of an extensive homebase for the entire community.

Led by Fan, with many family members cavorting about them, the swimmers eventually emerged into a sunlit lagoon, perhaps a half-acre in extent, with several inlets leading to small caves.

While they were shucking their dive gear, David commented to Kate, "There must be upwards of fifty CTs here if each of the caves holds a family. Could they all be the same family?"

"Sure," from Paul. "They're more of a clan than they are a family. Donna and I met a lot of them yesterday. Also I think they must have long lives because Tatty is at least a great-great-uncle. Not to mention a great-great-grandfather. They measure time by generations."

"And by warm and cold seasons," Donna added.

"I'll be damned," David muttered. "How d'you learn these things?"

Paul shrugged. "They show you. Well, I mean, he lined up a whole bunch of CTs, touched each one, and then put himself at the head . . ."

Meantime several cetasapiens were inviting the crew to explore.

"Stay in pairs," Kate instructed.

And was surprised when Frill endorsed, "Never alone in water. Always with others. Be safe," before leading them off.

But where Kate warned, Frill offered the reassurance of companionship.

Dave and Kate discovered that some of the smaller caves had specific purposes quite apart from creating quiet, private places for individuals or families. There was a general nursery, for instance, with several attendant elderly cetasapiens in charge of numerous young.

As they proceeded it became evident that cetasapiens enjoyed a communal life, sharing many activities.

Meeting back at the home lagoon, the humans donned their dive gear and were once again led by devious tunnels outside the main homebase, coming up in a cultivated reef area that Frill described as their home farm.

Most astonishing was a kind of museum, where Frill's kin arranged the objects garnered from shipwrecks. It was filled with all kinds of things that had provoked the curiosity of the cetasapiens, including bicycle wheels whose practical function apparently appealed as an art form.

A common cleaning station was crowded with CTs—and several other denizens—all lined up and waiting for tiny shrimp and wrasses to clean out their teeth, blowholes, and gills. A rushing conduit, off the cleaning station, was evidently used for waste disposal.

Here they grew various sponges and seaweeds needed for use in their healing lagoon, the promised destination for the next day. By the end of the tour, both CT and human language skills had much improved.

Back at the *Turtle* that night, Teiji, always map-crazy, created a very approximate cutaway of how the cetasapien volcanic island homebase might look.

The healing lagoon was a revelation: here the cetasapiens had utilized the naturally labyrinthine formations of volcanic rock to create quiet, open bays where various creatures as diverse as turtles and orcas and seals could be together without fear.

The crew was introduced to the head physician, who carried a spiny V-shaped crest, and to his many helpers.

As he led them around, expounding on his various patients, the aquanauts felt rather like a bunch of medical students being taken on "rounds." All of them were aware of the kinds of injuries he was talking about—cuts from propeller blades, patients in shock from being caught in nets, or infected by toxic wastes. But none had known that the sea creatures themselves were making desperate efforts to help the injured. Sadly, Dr. Vee reported that they could deal with only a very few of the thousands injured every day by human technology.

"Even you humans," he eyed them severely, "carry special knives to cut yourselves free of the nets, for instance. We do not have the dexterity to do so."

The group was silent. "However," he went on cheerfully, "we have other rather specialized methods for delicate manipulation when that is necessary, as, for instance, in an operation.

"But for the most part," he went on, just as David was about to ask a question, "our treatments consist of disinfecting, deadening pain—ah, re-setting dislocations, and generally providing a quiet, safe habitat that allows our sick ones to heal themselves. Now . . ."

"Doctor," David interrupted, "what about manipulation? You were saying . . .?"

"Ah, yes, yes, of course. Professor—ah—Schlessinger, isn't it? Well, let us proceed."

He led them over to a large area, which appeared to be a lab room, or possibly an operating theater. Here Dr. Vee approached an open, underwater tank housing several small octopi. To the dismay of the humans, he slipped into the tank, grasped one of the wriggling creatures and brought it out.

Everyone was relieved when he announced, "This little one is not for eating. These are specially trained to perform the tasks which we, with our clumsy thumbs, cannot do."

"So you put them on like a surgeon puts on gloves—sort of," Paul commented.

Dr. Vee looked bemused. Perhaps he was not familiar with surgery. But he nodded briskly and went on. "Exactly, young man, exactly." He seemed to feel he had a bright student here. "Symbiosis, you understand."

"Doctor, how do you control it to make it do what you need?" David wanted to know.

"Ah, now. These little ones, and their big cousins, the squid, are quite intelligent. They follow imaged directions, and respond also to charged impulses—nothing injurious, you understand—merely one of our standard methods of communication and treatment."

Kate thought of her ship in the grip of the giant squid, but kept quiet.

Dr. Vee kept the small creature with him, clicking disapprovingly as they approached a dolphin, still entangled in part of a net. Jay asked if he might help, and used his safety knife to cut the viciously strong polymer strands.

"We might be able to adapt these knives for your octopi to use," Teiji suggested.

"Yes, yes," Dr. Vee was enthusiastic. "It would need an—ah—abraded holder to avoid slippage. Can you manage that? And the blade would have to be very slender for them to be able to use it. Perhaps we could confer about this later?"

"By all means," David agreed.

Their next stop was evidently a birthing station, for it was clear that there were no sick creatures here. Instead, many small, busy babies, amazingly active considering how young they were, flitted back and forth between their strange visitors and their respective mothers.

Donna was enchanted. She tickled and played, obviously seeking close contact. She announced that she'd like to stay a while. Dave suspected that Donna had little stomach for the hurt and damaged animals they had been seeing, and readily agreed. He knew that beneath her tough exterior Donna had a heart of pure mush for small, vulnerable creatures.

They left her there, cradling a little fellow no more than eighteen inches long. He appeared to have fallen in love with Donna, or she with him. The parents hovered proudly nearby.

Jay in particular was fascinated by medical techniques, following Dr. Vee's comments closely, observing and questioning.

"I bet your 'X-ray' vision is handy," he remarked.

"Handy?" Dr. Vee seemed puzzled. Then, "Oh, I see. You mean useful. Yes. Both for adjustment and in diagnosis. By the way," he turned to David, "you are aware that you have a weakened ligament in your left knee, Professor?"

"I certainly do," David replied with some emphasis.

"Ah, in that case, might I be permitted—?" Dr. Vee was politely eager. Amused, David bared his "trick" knee, and shortly felt a warm glow spread through the joint. As he worked, Dr. Vee lectured the group, "We are using sonar here, to promote extra circulation to the affected area. Ligaments are difficult but by no means impervious to patience and treatment. Of course, any one of us can do this for you, Professor. It is a simple treatment."

Frill cheerfully undertook to give David a daily charge of sonar.

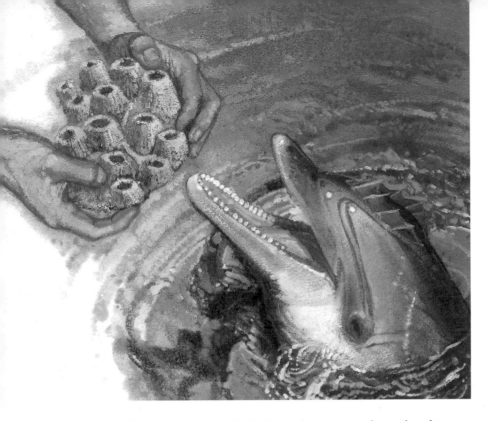

went into an elaborate explanation that included pressing and, if Jay understood him correctly, chewing. Although he couldn't make out exactly who, or what, did the chewing, he gathered the process provided an adhesive that held the dressing on in salt water.

Jay never talked much, being given more to action, but perhaps because his profession made him more of an observer, he noticed a great deal. He realized that the cetasapiens, at least between themselves, used an involved, multi-layered body language which the humans could not perform, except in limited fashion, but that it would be enormously useful to be able to understand it. He determined to do so, and thereby make his own contribution to communications.

Meantime he planned to spend time in the healing lagoon, watching all the other creatures, and learning new medical techniques.

During the several days that followed, one or two cetasapiens attached themselves to each of the aquanauts as mentors and guides around the homebase. Everywhere they were greeted with engaging curiosity and generous good humor. Between the body language, which they were all learning to interpret, the alphabet resonance system, CT imaging and Paul's music, communication rapidly improved between humans, CTs, and whales, of whom there were always two or three about in the big lagoon.

Kate wished they hadn't shucked their dive gear when she discovered their next encounter was with a squid, albeit a relatively small one, maybe only ten feet across.

"Be very careful," Dr. Vee instructed. "Octopi and squid are extremely shy and very easily frightened. And this one is very sick—from ingesting polluted molluscs. Oddly enough."

"Why 'oddly'?" Paul asked.

"Well, perhaps it's a mark of respect to savor one's ancestors, but definitely unwise if they are polluted." Kate wondered if Dr. Vee were making a joke, but he was continuing, "The cephalopods are evolved from molluscs, you know."

"Remarkable," Paul was nothing if not polite.

"Yes, indeed." The creature they were discussing could not have looked more miserable, head sagging, tentacles curled in distress. Kate decided this was a good opportunity to get rid of a troublesome prejudice, and putting out a tentative hand, she stroked gently. The squid immediately curled a tentacle around her wrist and turned a delicate pink.

"Ah!" Dr. Vee was delighted. "He likes that! Continue, please." Presently the tentacle relaxed. Kate was both relieved and triumphant.

Jay noticed that massage, combined with sonar, was in widespread use. The effects were obviously very relaxing. Lesions and wounds, however, required the application of a sticky ointment. He asked Dr. Vee about it, and got a voluble response.

"We extract a healing substance from a rare sponge—" here he

Paul spent a great deal of time with them and became absorbed in creating a musical instrument he could play underwater. Hours in the workshop with Teiji produced a hanging xylophone of copper tubing. Struck by a small hammer, this made sweetly plangent sounds that delighted the CTs, who excitedly set about learning how to play the instrument themselves.

Donna returned time and again to the homecave of the newborn she had first encountered in the healing lagoon when he was just hours old. The little fellow had formed a firm affection for her. Dave, Kate and Jay spent much time with Dr. Vee, who insisted that the squid's recovery depended on Kate's daily ministrations. Actually

he thoroughly enjoyed earnest discussions with David, and was flattered to respond to Jay's probing questions about cetasapien medicine.

From the homefarm, Dave and Jay collected specimens of painkillers, disinfectant plants, poultice weeds and other remedies to carry back to the lab. Each evening, back at the *Turtle*, the crew would go over the day's discoveries. As Frill had promised, they learned a great deal from seeing. All of them, but especially Paul and Donna, responded to the strongly family oriented community living in the busy homebase, one of many existing in the warm seas.

"You know, they're pretty vague on numbers, but I'd bet there are thousands of CTs," David commented. "The ocean is so huge: it's a truly vast, open water habitat. Despite their numbers and their obviously superior intelligence, they seem to be simply neutral participants. Apparently superior equals live and let live."

"The whales, too," said Paul. "They don't control or dominate."

Over dinner, the team had many talks, the burden of which added up to the gut question, why had the cetasapiens brought them here? What did they want of the humans?

Eventually David once again asked Frill for a discussion, and they all, the cetasapien community and the humans, gathered at the *Turtle*.

Kate opened the conference with a direct question that went to the heart of the humans' problem. "Frill, why bring us here if you have kept yourselves hidden for so long?"

"That is simple, my friends. Our way of life, where we all live, is threatened by humans, even if not intentionally. We needed to learn more about you than we had been able to glean from monitoring your electronic communications." Teiji's mouth opened, but Frill was continuing. "But we believe, even more, that you must learn about our world. To do so, we want to take you on a tour. The ocean is a very much larger part of our planet than the thin layer of land on which you live, yet you know so little about it."

"Frill," David protested, "that will take weeks. We've already been out of touch with the upper world for too long."

"That does not concern us so much as the possible worries your families might be suffering."

The humans looked at each other.

"I don't have any family," Kate said slowly. "Nor does Teiji. Jay?"

Jay shook his head. "No problem. My folks're used to me being away for extended periods."

Frill eyed David. "My family is here with me," smiling at Paul and Donna, "however, if we go, once we leave homebase we would be well advised to send a message that we need to conduct experiments requiring several uninterrupted weeks, possibly even months."

But Frill was watching Donna. "You are upset, little one. You don't want to go?"

Donna looked confused. "I want to go," she said, "but—I don't want to leave." She was suddenly close to tears. Support came unexpectedly from Paul. He put an arm around her shoulders.

"What Donna is afraid of," he said, "is that we won't ever see any of you again."

Frill was horrified. It was a possibility that simply had not occurred to any of the cetasapiens, and which the humans had not had time to think through.

"But of course we will have to return here," Fan said. "Many of us will go, some to bring back those we find needing help. Others will travel all the way, and in any case, we will eventually all return. This is homebase."

Uncle Tatty spoke up. "Some of us will stay here, Donna, to take care of the very little ones. You do not have to worry. We are all your family now."

Dave saw that, with characteristically gentle understanding, Uncle Tatty had put his finger on the problem. And it wasn't only Donna. Paul clearly shared her feelings, and somewhat to his surprise, he found himself deeply relieved to know that they would return. He glanced over at Kate. She nodded, and simultaneously they realized that in this crucial matter they had been consulted, rather than dictated to.

He rose and turned to the entire gathering. "Well, friends," he said, "I guess we're going to be together for quite a while."

The lagoon around them suddenly boiled and crashed with leaping cetasapiens. Through the uproar, Frill happily fluted, "Tomorrow we will have a real celebration!"

The next day, outside the homebase, where the farm reef dropped off into the deeps, the cetasapiens staged a complex ballet of dazzling speed and grace. Every so often one of them would engage an aquanaut in a minidance. At last, happily exhausted, all returned to the lagoon.

The *Turtle*, convoyed by dozens of cetasapiens, would eventually leave for their projected first stop—the Arctic, a journey that would take several weeks.

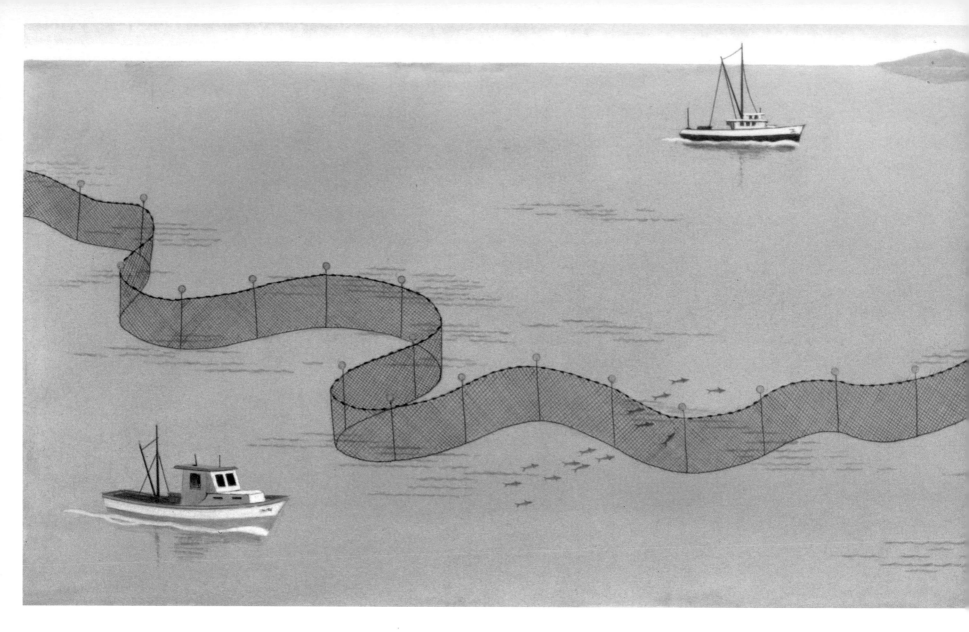

Their way north was impeded by an encounter with "ghost nets." Early one morning when Jay was on duty at the pilot station, Frill sent in a shrill alarm. Jay at once slowed the ship and roused Kate and Teiji. Shortly a very strange image appeared on the monitor. It looked like an endlessly long wall, thick with ensnared dead fish, sagging heavily in places, floating free in others. There appeared to be no way around it, although it was still quite some distance away.

But Frill was insistent. "We must avoid this," she said. "My mate and several others

are following the line to look for possible survivors among the turtles, dolphins and other captives, but they are experienced at dealing with this. He'll rejoin us later, and perhaps the others too if they don't find any survivors."

"But what is that thing?" Kate was aghast.

"These are the 'ghost nets' you were told about in dive school," Teiji said. "They are forty-mile nets outlawed some time ago and abandoned instead of being pulled in. They float and trap hundreds, thousands, of fish and other animals, sink to the bottom from the

weight, and when the dead rot out, rise up to do it all over again. The horror is hard to imagine until you come up against it. You'll be lucky if you never have to use your net knives, Kate."

"The nets're made of a particularly tough polymer," Jay's voice was grim. "You remember the dolphin we cut loose in the healing lagoon. I suppose there may be broken pieces here and there, but they are still one of the worst silent killers in the sea."

Frill agreed. "Pieces sometimes get snagged on bottom rocks," she said. "The cur-

rents eddy the broken filaments around into great, weaving traps. On the surface loose pieces get carried about so we never know exactly where we might run into them. When they are heavy with dead fish it is easier to hear them."

"You mean you have to run continuous echo-location?" Teiji asked.

"In certain areas, yes. But that's not as difficult as it sounds because we do it in relays."

"Like having scouts out," from Jay.

"Scouts for what?" came from behind them. "Why have we slowed down?" It was

David, always an early riser anyway, and curious about the change in routine. He was shortly followed by Paul. Kate explained the situation. The *Turtle* had backed off a good distance so Jay ran the image sequence they had picked up over again for their benefit. Presently Fan showed up, several others having returned to homebase with two injured dolphins.

In silence, the *Turtle* made a large circle and proceeded on its way.

Three weeks of travel brought them to the seamounts of the Emperor Range. The *Turtle* went down for a closer look at the sharply defined formations whose peaks thrust to about 2000 feet below the surface.

Donna, curious, accessed data on the computer. "It's part of the Pacific Circle of Fire," she announced. "Look at that! There are hundreds of active volcanoes down there. Even earthquakes."

"That's right," from David, "where the continental plates are moving and meeting and separating. Going on all the time."

"The sea floor is always heaving about?"

"Moving, building, shifting, changing. The magma comes pushing up through the rifts. Sometimes it'll push all the way up to the surface to make an island. Sometimes not. Some of the cones just keep building up and up and never reach the surface. That's why they form those peaks: they never get a real chance to blow off or get eroded from surface activity."

"Weird. A whole great mountain range moving, miles deep and hundreds of miles long, and nobody really notices."

"The cetasapiens do," Paul said.

Later that day Frill alerted them that they would again need to change course. A volcano was erupting close to the surface. The CTs had detected underwater seismic shock waves long before it could become dangerous to them. The sea creatures had evacuated the area.

Miles from the eruption the *Turtle* surfaced briefly, not wanting to miss the spectacular show. "Old Mother Earth giving birth to an island. How's that for a baby, Donna?" Paul teased as they all watched and listened in awe. "The noise from that thing must be horrendous under water."

But Kate became concerned for the welfare of her ship. "Dave, I think we'd better pull out of here: I know you'd like to stay and record the whole thing but—"

"No, you're right. There might well be a tsunami, and we'd better be down when that happens. Also if the winds turn our way, we'll have a problem with ash."

"Frill said to head due north," Teiji reported. "They figured to rejoin us in the morning." Which, in fact, the cetasapiens did, and the convoy proceeded to its distant destination.

When at last they reached the Arctic, it was obvious their arrival had been expected: every kind of far north creature was there to greet them, led by the big gray whales.

The vast ice-bound bay was alive with spouting, breaching cetaceans—orcas, grays, humpbacks, narwhals—the shores lined with aquatic mammals, the seals and sea lions keeping a wary eye on polar bear and orca alike. The aquanauts, able to enter the frigid waters only very briefly, watched much of the show on the big monitor.

Frill and her fellow cetasapiens were in a transport of excitement, darting back and forth, fluting, chirruping, leaping high out of the water, delighted to be participants in an orgy of reunions.

"So much for the so-called 'icy wastes,'" Paul remarked. "This place is positively bursting with life. And we can't even see the land animals, the foxes, hares, who knows what else."

Teiji said solemnly, "Yeah, it's a jungle out there," getting a laugh from everyone. Kate was peering out of the bubble, fascinated. "It's more like a market fair. I mean, it is positively crowded. Is it like this all the time? We must ask Frill."

*The tiny krill, base food of the Arctic, feeding by the billion on the spring bloom of plankton, attracts a rising chain of sea, land and air creatures.*

After a couple of days, when Frill had worked off the first explosion of her joy, she settled down long enough to talk with the *Turtle*-bound aquanauts—themselves the object of a lot of attention.

"Frill, why is there this huge gathering?" Kate asked.

"For feeding and breeding. We all come here to feed

"And why not? They have enormous power and grace. And they know how to put it to good use. They circle on the bottom, letting air out of their blowholes to create a cone of bubbles that trap and compress the krill. Then one of them surges up the middle of the cone, mouth wide open, and gulps in tons of krill. Very skillful. And beautiful. You will see."

off the krill. And each other. Little fish eat krill, big fish eat krill and little fish: seals, sea lions, dolphins—all eat big fish, and sometimes orcas and the polar bears eat seals. It is all part of the cycle of life. This is why we brought you here: the Arctic is one of the great life sources of the oceans, a creative whirlpool. Literally. As you will see from the bubble dance of the whales."

"Bubble dance?" Donna exclaimed. "You mean the whales dance?"

The bubble dance was all that Frill had promised. The aquanauts watched, enthralled, as the great mammals swirled, magnificent flukes pumping, every so often one of them shooting up the whirling bubble-trap. The torrent of sound from the powerful circling was like standing in the rush of a gigantic waterfall.

Later, when things had quieted down, a thoughtful Kate remarked to David, "You know, when we first arrived, they expected us. How did they know we were coming?"

"Well, I imagine the cetasapiens announced it. They're able to communicate with several different species."

Frill interjected, "But we are not as skilled as our large cousins. The whales are the true communicators of the sea. Under the right conditions of currents, temperature, and salinity, they can sometimes send information for what would be 5,000 human miles. They speak from sea to sea."

"Five thousand miles!" Paul was really excited. "How?"

"My friend, if I could answer that, we could probably do it ourselves. We understand some, but not all, of what the whales can do."

"Would it be something we could figure out together?"

"I doubt it, Paul. The whales are our wise ones. Besides being able to send sound-pictures over long, long distances, whales—" she thought for a moment, then went on, "they carry time."

"Now, wait a minute," from David. "Frill, you really have to be more specific than that. What d'you mean, they 'carry time'?"

"They are what Donna would call 'mainbrains.' They know the history of all our species—they carry time in their heads,

you see. The ocean is only a part of our world. It is world that they know. They—they think in whole-world terms. And once each year, they sing the song of earth, telling us all of the changes that have taken place, the differences in our terrain, the new volcanoes, the abundance of various food sources, what is happening at the interface with coastlines, and so on."

There was an awed silence. Then David said quietly, "No wonder they have such enormous brains."

As it turned out, the whales also had a lively sense of curiosity and were enchanted with Paul's music-making. Eventually they joined him in a joyous watery concert.

Luckily there were only maintenance chores to do on the *Turtle*, for the entire crew became dedicated Arctic-life watchers. The bubble deck was a favorite spot for viewing close-ups of the great variety of creatures. The tables were turned one day when the ship had moved out into the bay and a curious whale examined the inside of the command module.

Weeks went by as they cruised from bay to bay. Although they were not really equipped for Arctic land conditions, a couple of times they rose to the surface and either watched from the conning tower or actually got out on the ice. But eventually the *Turtle* got ready to leave with the grays on their annual 4000-mile migration down to Baja, off the California coast.

The grays generally followed the coastline, only occasionally veering out to sea. It was a leisurely trip, the pregnant females moving in family pods, leading other females and males, all heading for the safe, shallow breeding lagoons off Baja, where the cows would give birth, while the rest reveled in the warm waters and the joyous mating that would produce young the following year. By April or May, the young calves would be big enough to start the long trek back to the spring burgeoning of food sources in the Bering Sea and the Arctic. Half way to Baja, once again, big Fan left the group with his helpers to check the ghost nets. Flick and Fleck, already much larger than when they had

started out, stayed with the *Turtle* convoy. When they reached the San Diego basin off the California coast, Frill called a halt. She wanted to show the aquanauts the kelp forests. It was a priceless opportunity for a guided tour in forests where divers could easily get lost and even trapped in the sinewy 300-foot fronds. Frill said they could stay several days, making short dives in water that was still very cold for humans, and then easily catch up with the slow-moving grays. Although the *Turtle* would need to remain well below the surface, everyone welcomed the break in routine demanded by constant running. Their relief turned to delight with the first foray into the forests.

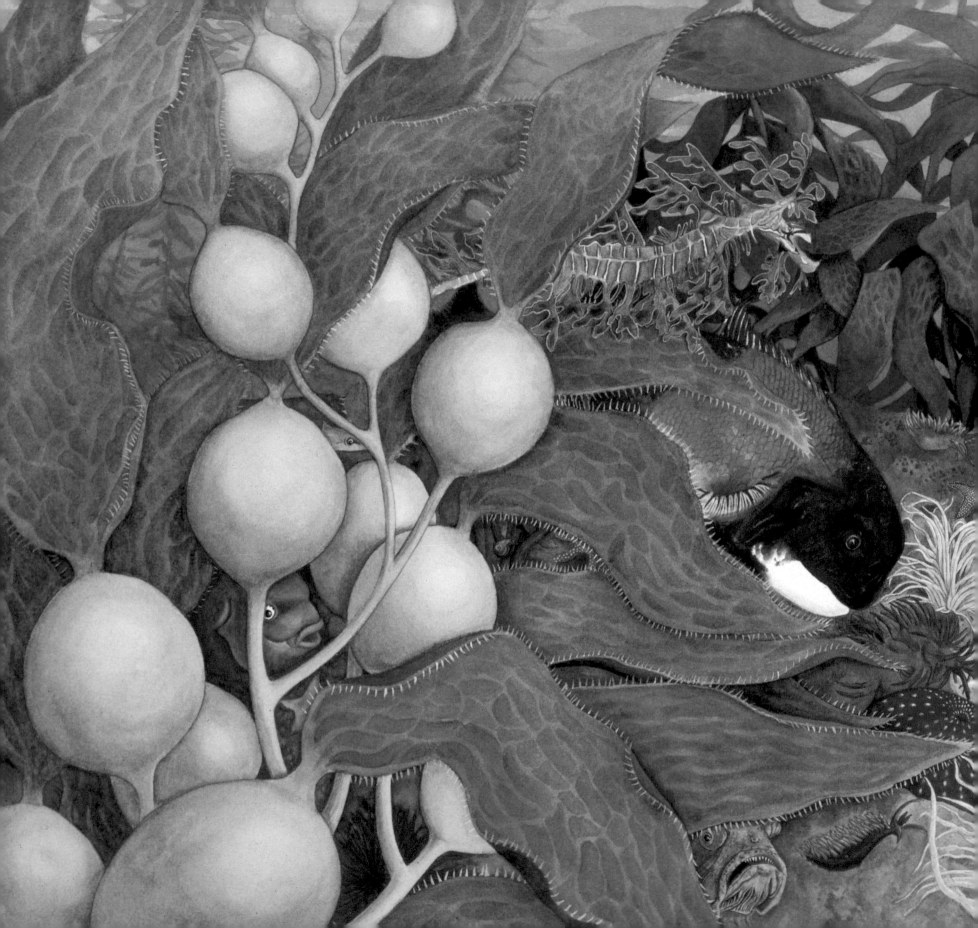

Dave was in his element, heading straight for the depths to show the others the root-like webs called "holdfasts" that anchored the constantly weaving kelp trees to the reef with "the strongest glue in the world."

The forests were immense, towering up a hundred feet and spreading another hundred each way in a dancing canopy, feeding off the sea and the sun. The great weight of the 300-foot growths was held aloft by gas-filled pods that grew from the base of each ferny frond, the whole supporting a myriad species safe in the bronze-gold gloom of its high-rise tiers—at least from large predators.

Richest of all because of the continual shedding of fronds, the bottom fairly coruscated with life—nudibranchs, anemones, lobsters, snails, bat-stars, octopi, eels, abalone—a host of bottom-feeders, including the voracious sea-urchins. This spiney menace ate the kelp itself. The forests had been seriously threatened when human hunting virtually wiped out the sea otter and the toothy-mouthed sheephead fish, both of which fed on sea-urchins. And only the sea otters, as Frill pointed out, were back in useful numbers. Helping to maintain balance were hordes of garibaldis. Although unable to actually eat urchins, the bold orange fish would pick them up and busily swim off to drop them in the sandy wastes outside the forest.

Kelp easily accommodated the minute chomping of the tiny Norris snail, which ate its way to the top of the sinewy Everest, only to launch itself off and start all over again. Nor did other climbers and eaters, the crabs and molluscs, bother the rampant growth. Great storms could sometimes tear the kelp loose, but it would always grow back, for the giant ocean stalks thrived in cool coastal waters.

But warming seas would cause the forests to wilt, and die.

It was hard for Kate, Paul and Donna, seeing this swaying cradle of life for the first time, to believe that it could be destroyed by anything. Shoals of mackerel, yellowtail, bonita, and many other fish patrolled the perimeter. Within the central forest, countless unnamed beauties flicked and fled among the fronds.

One variety, the kelpfish, actually sported frond-like growths to imitate its host. This clever disguise allowed it to hide from predators and to pounce on its own unsuspecting prey.

No one had any trouble avoiding the six-foot moray eels, sprouting impossible teeth and lurking hopefully in their reef nooks to grab unwary fish or octopi. The aquanauts watched while one potential victim emitted a cloud of ink that deadened the eel's sense of smell.

Dave said the scorpion-fish was actually much more dangerous than the eel. He carefully pointed out the foot-long venemous creatures almost totally concealed in reef rocks, and looking like just so many sand-covered rocks themselves.

On the surface, the kelp spread its great fronds to catch the sun, creating a gently rocking bed for sea otters that periodically dove to harvest clams and abalone. Lying curled in the comfort of the cradling waterbed, otters could crack the clamshells, using their chests and stomachs as a handy worktable. During the day harbor seals basked on the reef, soaking up sun.

Above them, birds screamed and dove for fish, and below, the incredibly swift and graceful sea lions played tag among the weaving stalks, constantly rehearsing for the game of life.

The sea lions were endlessly curious, frequently approaching the aquanauts to investigate this strange life-form. They were also apparently fearless, racing out to the edge of the forest to tease and nip at the makos, blues and even the occasional great white that cruised by attracted by the banquet the forests provided. Kate disapproved of the sea lion bravado. Frill was amused.

"It is far more dangerous to show fear," she said, "better to practice dexterity in escape, or confuse he who threatens."

"Or be careful in the first place," Kate responded.

Each morning the aquanauts were up early, anxious to get to dive sites so that all could manage two dives a day. Jay and Dave, as the most experienced, usually led separate expeditions, always accompanied by several cetasapiens. Evenings, Frill or Fan would have liked to lecture, but quickly found their human audience had a tendency to drop off to sleep. They settled for music and murmured talk.

"Have you ever seen the summer mating of the bat rays, David?" Frill asked.

"Yes," David replied, "spectacular at night. Swirling around like flocks of great, graceful birds in the moonshot clearings."

"The kelp is a breeding ground for many other species that come here every year to renew the cycle of life. Soon now the squid will come in their billions," she went on with relish, and then, seeing Kate's startled look, Frill chuckled. "They are very small, Kate. They mate and blanket the sands with their eggs in a layer two or three feet deep. Then the parents die, and all the ocean's creatures come to gorge on the crop of squid. You will see."

And during the days that followed, from the safety of the *Turtle* they watched every variety of shark gulping down hundreds of dead squid, joined in the orgy by sea lions, harbor seals, crustaceans, dolphins and even a whale. The detritus of their leavings fed thousands of small scavengers. In this gluttonous feast, the egg sacs were ignored, left to mature and drift away. In one year, Frill said, they'd be back as full-grown squid to perform the same ritual.

Great Hammerhead Shark

Great White Shark

Some of the bottom-feeding sharks were year-round visitors, able to hunt the sandy wastes by sensing the tiny electric impulses emitted by fish that had taken false refuge by burying themselves there.

Blue Shark

Sand Tiger Shark

Angel Shark

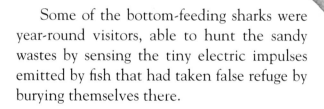

The idyllic stay in the kelp forests was brought to an abrupt halt one day when Frill, who was with Donna, David and Teiji, suddenly became very agitated. She urged them back to the ship, to find Fan there with Jay, his group having already entered the lock. There were not many cetasapiens about, and all seemed uneasy. This was so unusual that both David and Kate became alarmed.

"Frill, what is it? What's happening?" Kate wanted to know.

"There's a whale in trouble down the coast," Frill replied. "We must get there as soon as possible. Get everyone into the ship. We have to leave at once."

"All right, of course. But we need a little time to prepare for departure, Frill. We just can't leave immediately."

"Then I'll go ahead. I'll leave three others to guide you. Flick and Fleck, you stay with the ship and your uncle and aunts. Keep together. Be safe." And she and Fan were gone with a flip of their powerful flukes.

While they waited as, one by one, the aquanauts got through the lock chamber, David consulted with the remaining cetasapiens. "Can you tell us exactly what is wrong, and where we'll be headed?"

The male CT, bearing a spiky circular crest, was weaving restlessly. He paused to reply. "One of the mothers is giving premature birth out in the open sea. Most are safely in the bay but she left late. This happens sometimes, and all such calvings are an invitation to every passing shark for miles around." The very idea sent him dashing back and forth.

One of the "aunts" continued, "Normally we would not interfere. It is in the natural order of things for premature calves to be lost. But things have not been natural for some time now. The whales are few, and growing fewer. Each one is precious so we must do what we can to help. Hurry."

Dave and Kate shortly entered the ship and found the others already had preparations under way. As soon as they were mobile, the cetasapiens led them out and then arrowed south. The noise of the *Turtle*'s thrusters at full bore must have been painful to them but they kept well ahead. Teiji and Donna took the first watch, relieved in a couple of hours by Kate and David, and they in turn by Paul and Jay.

The *Turtle* raced through the night and just before dawn sonar soundings told them their guides were slowing down. They followed suit, and as dawn broke, drifted in, just in time to see the calf being born.

Shortly after its first breath the calf had begun to feed. Things were so calm that Kate and Jay actually entered the water.
Moments later the sharks appeared.

Teiji, watching from the ship, realized there were too many sharks for the cetasapiens alone to cope with. Kate and Jay were safe enough near the lock, but the threat to the others had to be stopped. And fast.

Hands moving like lightning over his keyboard, Teiji sent a warning to the CTs and focussed a narrow sonic beam at the approaching marauders. There was a violent reaction, and most of the sharks disappeared rapidly back into the depths. But a few, thoroughly disoriented by the blast, drifted helplessly. Teiji felt sick. He had not intended to kill them, but they would certainly die if they could not move.

Then, to his great relief, the cetasapiens swirled in, supporting the beasts until they recovered enough to move groggily away.

The annual birthing of calves at Baja attracted many watchers so the *Turtle* did not enter the lagoon but stayed at sea, hovering subsurface, out of sight of humans, but very much within hearing of the whales. Each day one or two males stopped by to inspect and thank the rescuer of the calf, whom they clearly identified as Teiji. Although none of them lingered, reluctant to be absent from the excitement and joy erupting in the lagoon, the aquanauts often made shallow dives to enjoy the company of the great, gentle mammals.

Then one day the whales brought the news that the Amazon was flooding. Frill and the other cetasapiens were elated. "This is a very important event," she explained. "We don't get there every year but I particularly want you to experience it. We will need to leave soon."

"Small problem," Jay pointed out. "Even though the Canal was modernized before turning it over to the Panamanians, how do we get through without being spotted?"

That really had them stumped for a while. Teiji fumed, afraid for the ship. Frill fretted, for once impatient. Then the heavy thumping of a passing oil-tanker offered a high-risk solution, in which timing and position were of the essence. Figured out first on a chart, the *Turtle's* sonar equipment and the cetasapien echo-location senses worked splendidly together.

Only one very puzzled operator at one of the gates ever wondered about the weird noises the passing ship seemed to be making . . . .

When they arrived at the Amazon delta, the aquanauts began to appreciate why Frill and the others were so anxious to have them see the phenomenon of its annual six-month-long flooding. The waters had risen a full three stories, and spread the delta 100 miles across, an expanse of yellow-brown channels overshadowed by livid jungle growth.

The unique trees and creatures of the Amazon had long since adapted to being waterlogged for half the year—indeed through eons of time fish had evolved special teeth to eat the fruit that dropped from the inundated trees. Many of these appeared to have been killed by the water, lifting ghostly white branches in the breathlessly hot and humid air. But they would recover, the cetasapiens reassured the humans, and meantime, the nutrient enriched flood spawned vast numbers and varieties of highly specialized fish, animals and jungle growth.

But Dave complained that they couldn't really see anything in the muddy flood, so Teiji carefully maneuvered the *Turtle* into a well-concealed channel from which the fascinated crew could observe the myriad life-forms. Brilliantly colored tree frogs, lizards and macaws leapt and crept and flew, vying with monkeys of several sorts. Caymans, capybaras, anacondas and just about everything else appeared to be virtually amphibious, while fish jumped out of the water and cormorants dove in. The crew settled in happily for a stay of at least several days.

Frill felt safe in calling her relatives, the Amazon dolphins, to this watery hideout. And another reason for the cetasapiens' eagerness to take the risk of bringing the *Turtle* here became clear. For the Amazon dolphin was a very special creature, having an extremely long snout—actually a sensory-perception device from which, as Frill pointed out, their own sensory head-crests had evolved. Moreover, the bone structure of this pink-bellied dolphin was the most primitive of all the dolphins, making them possibly the link between humankind and cetasapien. At any rate, there was no doubt in Frill's mind that these were the ancestors of her race.

Peering into the turbid waters to find the rather shy creatures, Jay remarked to Dave that it was no wonder they were not seen too often.

"They're not seen because there are so few of them," Frill pointed out. Jay felt a stab of guilt.

"You mean because they've been hunted out?" he asked.

"Partly that. Partly that the river is changing. They tell me that upriver a lot of the jungle is disappearing. The delta is bound to feel it eventually."

Jay slid gracefully to a related subject. He was, in fact, concerned about his cetasapien friends.

"Uh, Frill? How about that muck you're in? Is it difficult for you?"

"No problem," came the cheerful reply. "We don't bother with water breathing while we're here. Though to tell the truth, it's not much fun. We prefer the clean open sea, but our ancestors never leave the Amazon, so we have to come visit them here. They're quite happy with this glop, and of course it's exactly what we need out in the deep water."

"How come?" came a puzzled query from Donna.

"This is the stew of minerals and other nutrients that feed the sea year round—millions of tons of it, Donna." Dr. Vee was always pleased to advise the young human female he had virtually adopted. "The connection between our oceans and all large rivers is the connection of life. Dead rivers would mean a barren ocean. The rivers, among other sources, serve to nourish the seas. The bounty of the ocean bottoms," he went happily on, "is matched by the riches of the land. This is why all coastlines and swamplands are so vital. Life in all its diversity flourishes wherever water and land meet, creating and sharing a mutual feast." Dr. Vee was quite carried away.

"Speaking of food," Teiji remarked, "our supplies are low, and the ship needs servicing, also." With Teiji the ship might not actually come first, but she was at least an equal partner. . . .

Dave joined the discussion. "I hate to leave," he complained wistfully, "there's so much to learn here, so much to see." They all grinned at one another and a curly wave appeared on the screen—Frill was chuckling.

"Dave, there'll always be more to see than we have time for. Food and rest are also needed. Perhaps we'd better think about returning. On the way we can find an island suitable for servicing both the ship *and* its people."

They all left the Amazon with regret, despite the steamy climate. Dave and Jay in particular vowed to get back some day. But Teiji was really worried about the state of his ship.

"You're right, Taje," Paul said solemnly, "she's probably grown fins in that brackish stew."

"Then you'll have a lot of fun cleaning them off," Teiji said without batting an eye.

However, Frill found them a lovely little island, remote and isolated, with coconuts and even some bananas (Dave said it must have been occupied at one time), with a bay in which the cetasapiens found seaweed they claimed was edible for humans. Jay, a man of many talents, devised some acceptable way of preparing it. The sea was full of fish, just waiting to be caught, and the CTs delighted in bringing home a daily catch. The fresh food was a welcome change from reconstituted rations which had become both meager and monotonous.

There wasn't much to explore, at least on land, but everyone felt they'd had enough new experiences for the time being, and were glad to relax into the familiar. Paul and Donna spent a lot of time with the cetasapiens in the water.

Mornings the crew worked on the *Turtle*, anchored well out. Afternoons were dedicated to just plain lazing around or gathering food for the evening meal, usually cooked on a beach fire; from the sea, or comfortably lounging in rockpools, the cetasapiens watched and joined in the music that inevitably followed the picnic meals.

Several times the crew made night dives out beyond the *Turtle*, always with their cetasapien family, always with wonder at the breathtaking beauty the lights revealed. They had adapted the handholds on the lamps so that the CTs could carry them as well. But many nights they simply stayed on shore, bunked down on the sand.

It was a time that made a gift to human and cetasapien alike, a gift of sharing that wove a bond of joy between them, a magic time.

All too soon, it seemed, they were rested and ready to set out on further adventures. But Kate was worried about supplies. Frill knew her human family needed to restock with nonperishable foods. As soon as they were back in Pacific waters, they dropped down the coast of Peru to a spot where the whales had reported a recently sunken ship, an ancient freighter that had foundered and lodged on the continental shelf about a hundred feet down. She had been a supply ship loaded with dried and sealed foods, compact, lightweight, and hence ideal for transfer to the *Turtle*. Flick and Fleck regarded the operation as a game, busily shuttling back and forth. Jay hung packages on their "beaks". They spent the rest of the day expertly tossing, swapping, and catching their new-found toys.

Loaded at last, they set off for the rampart of the East Pacific Rise that runs from north to south for thousands of miles, marking a vast undersea continent of towering mountains, great crevasses, depths that dwarf the Grand Canyon, fierce currents and volcanic upheavals that are continually changing the ocean floor.

Approaching the East Pacific Rise, Kate brought them right up against a cliff that rose several thousand feet straight from the ocean floor, which, at that point, was some 12,000 feet deep.

The cetasapien contingent followed them to 8,000 feet and then went up to feed and rest while the *Turtle* explored. All the way down there were bioluminescent life-forms. Dave was so fascinated he went twenty-four hours without a break until Kate protested that he'd be useless on watch.

They did not again descend to any great depth for the next two weeks of the long journey across the undersea continent, headed for the Great Barrier Reef.

Frill was apologetic. "If we could go far enough down we could all take advantage of the undersea rivers, provided they were flowing in the right direction, of course," she said. "We can sense effects from the bottom but we can't take the pressures. Actually, the bursting fires would be dangerous for all of us."

Umbrella Mouth Gulper Eel

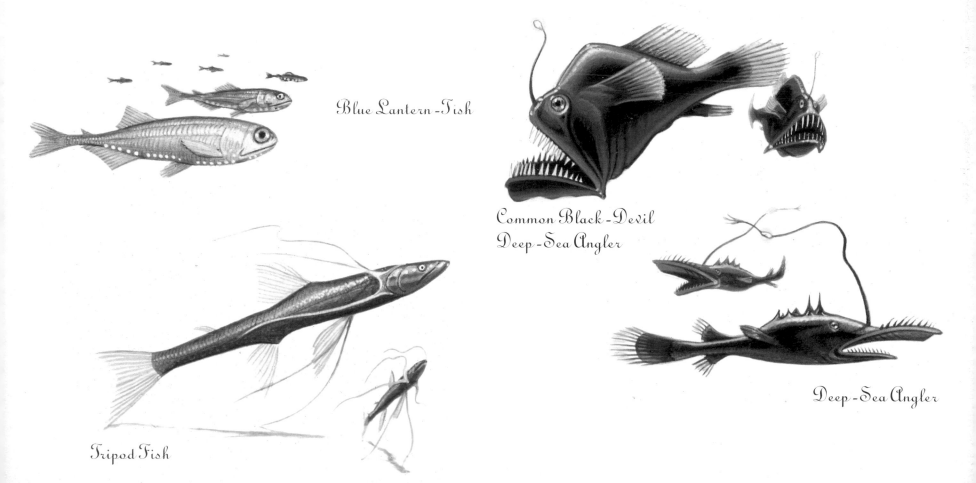

Blue Lantern-Fish

Common Black-Devil
Deep-Sea Angler

Deep-Sea Angler

Tripod Fish

The aquanauts were philosophical about it. They were all familiar with the phenomenon of currents created by changes in temperature and pressure, combined with volcanic and thermal activity. And Donna was delighted to find the cetasapiens eager to study information that she could access. The CTs, with the help of the whales, had long since charted the volcanic areas that floored their ocean home, but had never actually been able to visualize the cataclysmic changes and shifts that occurred when magma forced its way up through the crust in great rifts that spread layers of "pillow" lava miles in extent.

At the rest periods, the cetasapiens clustered around the bubble, full of questions about the charts that Donna or Jay put up on the monitor. They particularly liked the cutaway images as these came a little closer to the 3-D form their own senses provided when they did have close visual contact.

When they sensed a vent system, however, Dave insisted that the aquanauts must stop to investigate. Frill couldn't have been more enthusiastic, having an enormous curiosity about this unseen part of her world.

Dave and Jay went down in the shuttle. They flew among dramatic towers, created when seawater seeped down through the ocean crust, hit the magma, and spewed up superheated gouts of black, mineral-heavy steam, bursting into the icy seas at temperatures as high as 750 degrees, and building up chimneys as high as 200 feet. These were the cones known as "smokers."

Yet life abounded at these fantastic temperatures, and even on the chimneys themselves. Indeed, life clustered around the superhot water— strange life adapted to the chemicals leached out of the mineralized steam, and created by the heat of the earth's core rather than by sunlight.

Back at the *Turtle*, with the cetasapiens crowded around the bubble watching the video of the excursion, Dave told them, "There are theories that these hot vents might even be the original source of life on earth."

"Hmm." Dr. Vee was doubtful. "Although," he conceded, "they occur throughout the oceans and particularly in relation to volcanic activity.

"However, we have no trace in our history of any adaptation to a chemical synthesis-based life-form. So it's a far-fetched theory. But of course, as no doubt you are all beginning to realize, our world is capable of producing an infinity of life-forms. For instance . . ."

"Dr. Vee, could we see the video again?" from Donna.

"Of course, my dear, of course. Ah, thank you."

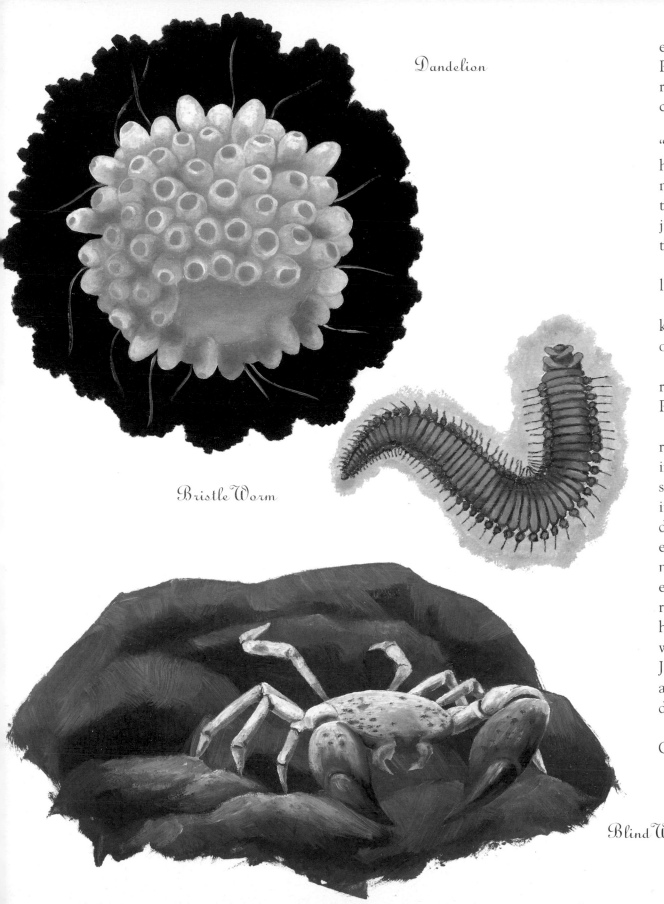

Dandelion

Bristle Worm

Blind White Brachyuran Crab

The idea of a life-form based on an entirely different chemistry had engaged Paul's imagination. As they all watched the re-run he asked Frill, "D'you suppose those creatures can hear?"

"Of course!" Dr. Vee promptly responded. "Well, perhaps not in the limited way that humans hear, and not even by the same means as we do, but they certainly sense things, Paul. Logic dictates that they cannot just float around waiting for food to fall into their mouths, or maws, or claws, etcetera."

"What about the ones that are anchored, like those nine-foot worms?"

"Ah well, yes. Of course. Detritus, as you know, is a vast food source for many of the ocean's creatures. Yes, indeed."

Everyone ignored the fact that Dr. Vee's response had nothing whatever to do with Paul's question.

The next day they set off again. By the route Frill selected, in a week they were well into the Coral Sea, following a relatively shallow ridge, when they were hailed by an immense blue whale. The aquanauts were delighted to get into the water again, greatly enjoying their encounter with this large member of the whale clan. After he had exchanged news with Frill, he courteously requested musical entertainment! He had heard of Paul's talents through the whale net-work, and Paul, accompanied by Donna and Jay, was delighted to oblige. Shortly there-after the great blue made his farewells and disappeared effortlessly into the depths.

The next stop for the *Turtle* would be the Great Barrier Reef.

On the final leg of the journey to the Reef, David set up a series of lectures about the nature of coral. The cetasapiens were enchanted to see their world in magnified form on the screen, and gathered eagerly around the bubble, fascinated both by human technology and the results it produced.

"In talking about coral," David began, "it is always difficult to decide whether to concentrate on the magnificence and the stupendous size of the structures built by nature, or to examine the delicate minutiae, the tininess and incredible variety of life coral cities and castles support."

"I go for the small stuff," Jay said. "I can hang over a few square feet of coral for hours, and not run out of new things to find."

"Right. So let's turn on the monitor and look at the marvel that cannot be seen by the naked eye.

"First of all, coral is not stone. It is a living animal that combines with vegetable matter—algae—to help build on itself. A part of it is made up of polyps, a kind of soft skin coating its calcified ancestors. These polyps are basically a feeding mechanism. The polyp has a mouth, with tiny tentacles—you can see them clearly in the blow-up— that gather food."

"They look like chrysanthemums, or like anemones," Donna remarked.

"Right. Corals are related to anemones, whose tentacles also trap food, and are sometimes poisonous, like many polyps.

"With coral, the polyp passes food to its partner, the algae, which acts as a kind of digestive system to nourish the polyp. Reef-building occurs when the polyp exudes a protective skeleton of limestone, to create a little hard cup. As the reef grows the algae are carried closer to the sunlight, which is what *they* need to survive and reproduce. So it's a self-creating process."

"How do reefs get started?" Paul wanted to know.

"Well, eventually the self-creating process joins with other similar growths to make large colonies. There are other reproductive methods. One variety of coral periodically spews out clouds of spores that get carried by ocean currents. And some find suitable homes, to found new colonies. Not all corals build reefs. The soft corals have brilliant colors of every variety and hue and grow every which way. They often anchor on the hard, reef-building kind.

"And so we come to the magnificence. It takes millennia to build a reef, especially the barrier reefs. The Great Barrier, for instance, runs for 1,200 miles. In fact, if you could link all the hard coral reefs of the world together, they'd cover an area twenty-five times bigger than the United States, which also says a lot about the area of the ocean world."

Dave's remark about the extent of coral reefs caused a stir even among the cetasapiens.

He went on, "There are four major varieties of reef. The barriers are the big one. Then there are the atoll reefs formed around a lagoon where a volcanic island might have collapsed. . . ."

"Similar to our homebase reefs," Dr. Vee pointed out.

"Exactly," Dave agreed. "And smaller, fringing reefs, along with the really small patch reefs. But of whatever size, they all support an incredible variety of life-forms. And of whatever kind, coral reefs are essential to the oceans."

"It is so," Frill confirmed, adding her favorite phrase, "you will see." Then she pointed out gently, "They are also essential to the welfare of the land."

As they got closer to the Great Barrier, Frill warned that the *Turtle* would need to remain outside the main reef, and even there would require concealment. Inside the reef, she reminded them, the waters were crowded with vessels of all descriptions, bearing sight-seeing tourists and divers.

Dr. Vee, already in a frenzy of excitement at the prospect of harvesting his sponges, was devastated. "They only grow inside the reef," he protested.

But Frill was adamant.

She was worried too, that the weight of the *Turtle* could damage the reef, and it took quite some time of quiet cruising around to find a suitable site, as well as very delicate maneuvering from Teiji to settle the ship. But eventually they were tucked in.

Above them Frill and Fan and the other cetasapiens organized thousands of schooling fish to act as temporary cover while the aquanauts anchored drifts of loose seaweed over the *Turtle*, blending her with her background in the best oceanic tradition.

All of them were, by now, personally involved with the fascination of life under the sea's surface. And all of them, in the days that followed would once again be seized with fresh wonder at the sheer variety of beauty and diversity that burgeoned in the sunny waters of a coral reef.

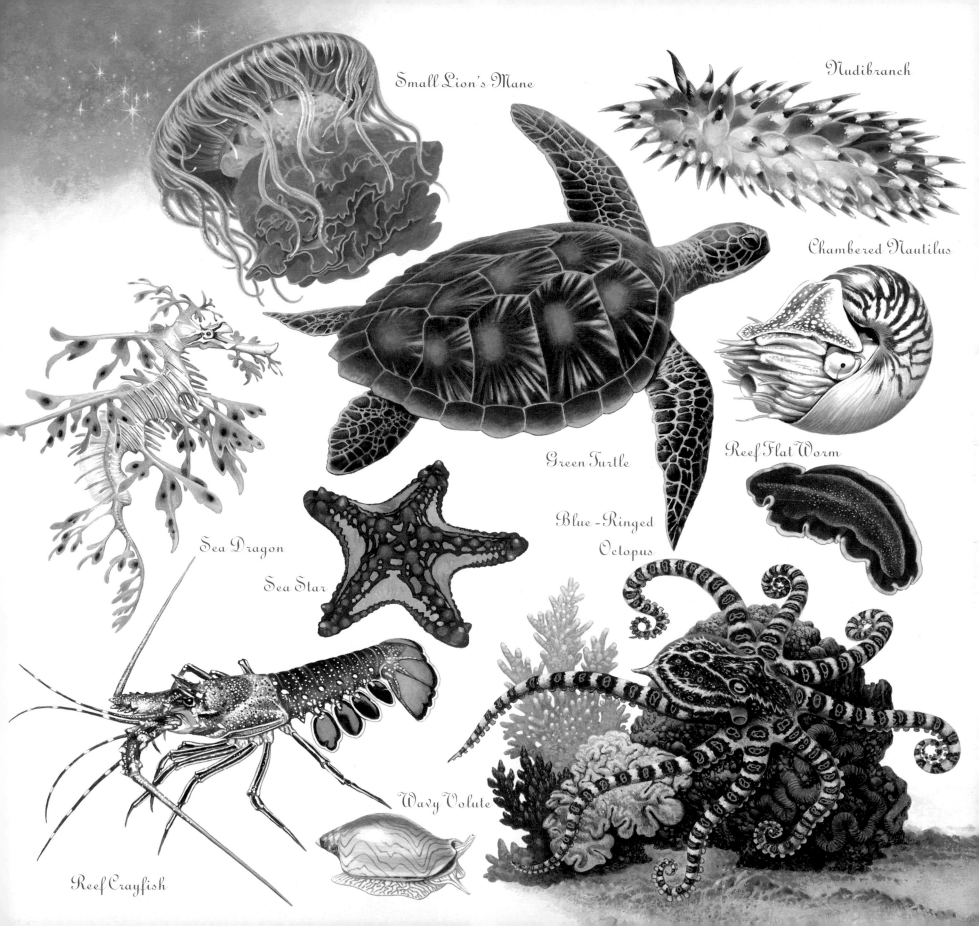

Small Lion's Mane

Nudibranch

Chambered Nautilus

Green Turtle

Reef Flat Worm

Sea Dragon

Sea Star

Blue-Ringed Octopus

Wavy Volute

Reef Crayfish

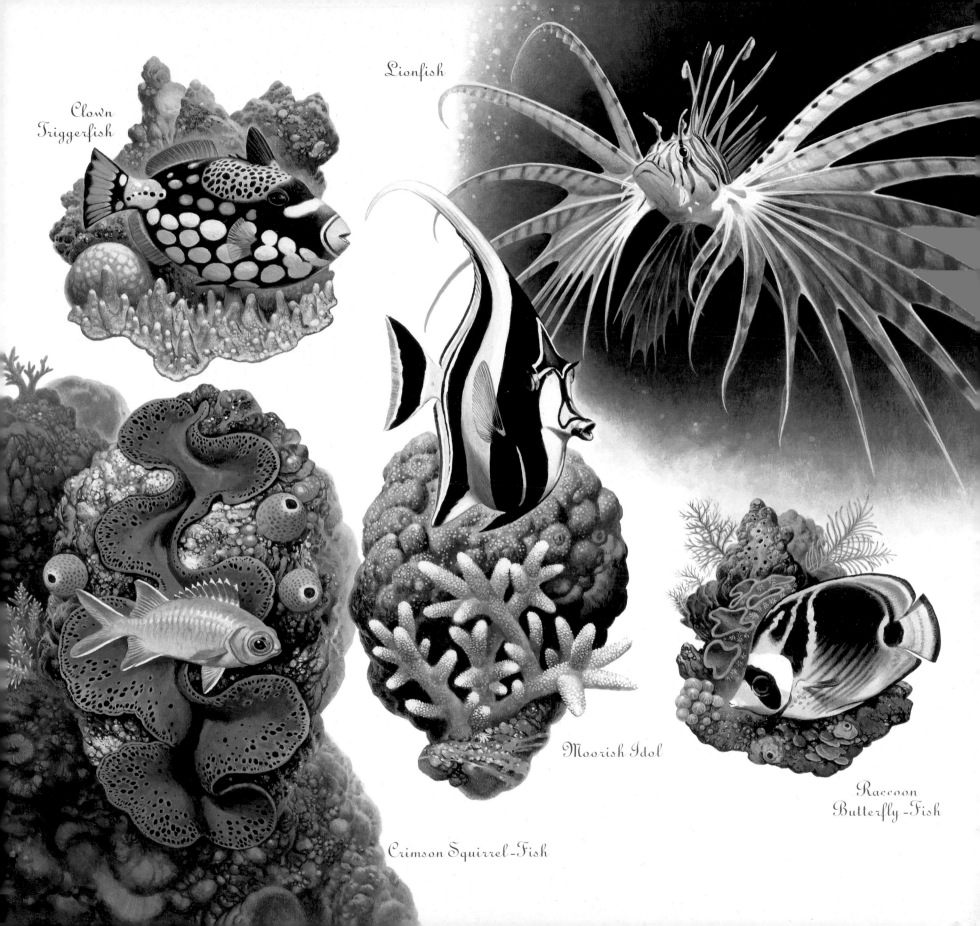

Clown
Triggerfish

Lionfish

Moorish Idol

Crimson Squirrel-Fish

Raccoon
Butterfly-Fish

During the many dives that the aquanauts made in the following days, all searched diligently for the sponges Dr. Vee wanted. But they were hampered by the need to remain concealed. Discussing the problem one evening, Jay said, "Hey, why don't we make a night dive? I'd like to anyway."

Dave heartily agreed. Busy recording all the new varieties of anemones, copepods, crinoids and countless other denizens of the reef that they encountered, he welcomed the peace and quiet a night dive offered.

They made plans carefully, each aquanaut accompanied by a cetasapien. It was Donna's turn to be left on board to monitor the *Turtle*. Unless they were anchored in a safe bay, the ship always had at least one person aboard.

To cover more ground the divers spread out, Fan and Paul paired together. They found a likely-looking crevice that gave the big cetasapien a little trouble but he managed to eel his way through. Inside they were enthralled by the explosion of color their light revealed. The cave was a wonder of vivid reds, purples and flaring yellows. Paul was mesmerized. When at last they turned to leave, they realized a large number of sharks had gathered at the entrance, and moreover, were very excited.

"Fan, I don't think I can get through that lot," Paul said quietly.

"Quite right." Fan was brisk. "Something has upset them—maybe the lights. I'm going to move forward a bit and get help."

"Wait a minute, Fan." Paul was not about to let Fan put his body between himself and the threat outside. "The *Turtle's* too far away to do anything."

"I know. But Paul," patiently, "I am better equipped to handle this than you are. Put out the light and be ready to come the instant I call."

Reluctantly, Paul had to agree. And in the pitch darkness, he waited.

Fan's solution to the problem was not one that would have occurred to Paul.

After what seemed an interminable time, Paul got an urgent beep on his wristcom, and having carefully oriented himself was able to move toward Fan. Beyond Fan, at the entrance, he found two large manta rays undulating gently, one above the other. Fan indicated the two remoras suckered to the back of the lower manta, and Paul understood at once that he was expected to hold on to them for a free ride. It was both exhilarating and hair-raising.

Paul clung to the remoras and the mantas sailed through the encircling sharks, an unnerving maneuver, even though the attention of the predators had been additionally distracted by a couple of dozen cetasapiens who darted among them, sending carefully directed bursts of resonance to confuse the would-be attackers.

Back at the *Turtle*, Kate demanded an explanation. Everyone had been alerted to get back on board the instant Fan's call to the cetasapiens had gone out. Normally the aquanauts would have gone to the aid of any diver in trouble, but none knew who it was, or precisely where. Besides, Frill's instruction had been clear, urgent and insistent.

Now it was Kate's turn. "All right, Paul. Let's hear what happened."

Paul was embarrassed. "Well, it was stupid, I guess. We let ourselves get caught in a cave by a bunch of crazed sharks.

"Fan blocked the entrance and called for help, and as soon as it came, I got out. He was terrific." He stopped abruptly. The adrenaline rush over, what might have happened had suddenly hit home.

Kate eyed him and commented, "I don't doubt it." Privately she wondered to Dave why Fan had allowed such a situation to develop.

"He's big, and smart, and strong," Dave said. "He doesn't think of himself as ever being in danger, at least not from sharks. And actually of course *he* wasn't. But that was one hell of a solution he dreamed up. I must try it sometime . . . ."

After the melodramatics of Paul's rescue, the members of the expedition had a serious discussion about the search for sponges. Night dives did not offer enough visibility to find them. Besides, Fan and Dr. Vee agreed that although just a few would be needed, they were convinced these were obtainable only inside the reef.

"Well, if we must move inside, I could have the Bug go in and collect them," Teiji offered.

"No, no. We can't afford to damage the reef," the doctor worried.

"No problem. I can manipulate the Bug to be as gentle as a mother," Teiji grinned.

"Truly? You people do have very useful machines. You don't suppose *I* could—?" Teiji wiggled ten fingers. "No," the doctor shook his head regretfully, "I don't have the right appendages. Well, never mind,

my friends. If you can do it, let's get the search organized."

The next day the team set out. Teiji was delighted to get some practice with the robot probe, and neatly steered the Bug through the reef channels. Even on the screen Dr. Vee's excitement was visible.

Once inside the main barrier, the cetasapiens hovered over the little Bug, effectively obscuring it from the boats riding above, while the Bug delicately collected the sponges they pointed out. Each then carried one sponge back to the waiting *Turtle*.

Back at the *Turtle*, Jay was at the lock chamber, ready to collect the sponges and pass them on up to David. Dave explained the workings of the specially adapted hydroponic tanks to Dr. Vee and Frill. Frill was most impressed.

"You could bring one of us aboard to live in a tank," she remarked.

Dave hesitated. "Well, yes," he conceded, "we could. But it would not be very comfortable."

Dr. Vee gave the equivalent of a snort. "We are scientists, I hope. Who thinks of comfort when it comes to the discovery of knowledge? It would be a most interesting experiment, professor."

Dave was actually delighted with the idea but aware that his friends might not quite appreciate the discomfort involved. "Well, all right, we'll think about it. But let's get one thing done at a time. Jay and I will get these sponges bedded into their holding tanks. Then we all need to decide where we'll be headed next."

And in fact, within a few days, the *Turtle* did indeed set out north, with a planned short stop at the Mariana Trench. They were diverted momentarily by a Fin whale from the Southern ocean, a loner who nevertheless was a member of the network, who informed them that their progress was being followed with interest all over the oceans!

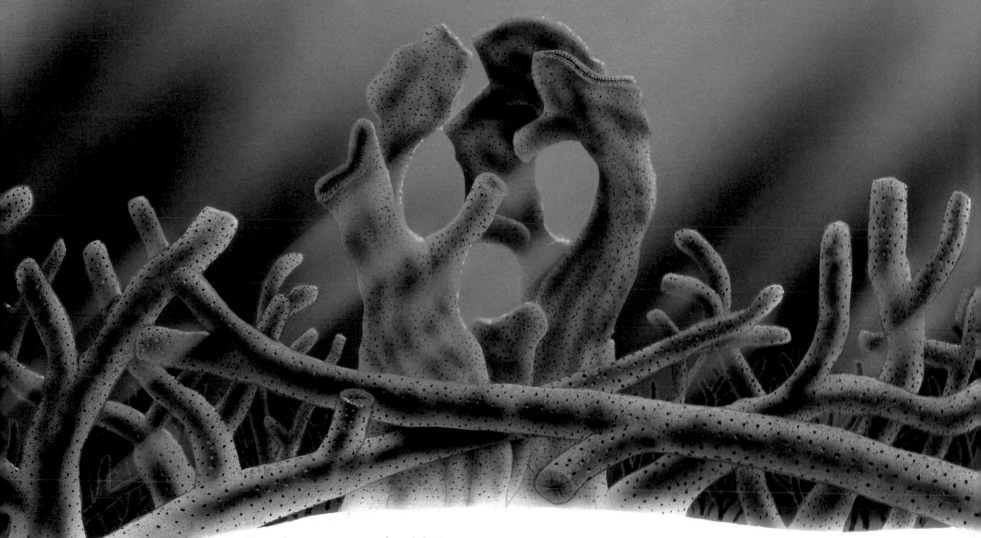

On their way to the Marianas, relaxing in the saloon with Donna and Teiji, Paul wondered if they mightn't make a sidetrip to the Philippine reefs. "Near your old home. Right, Taje?"

The usually ebullient Teiji was silent. Then he said abruptly, "No point."

When Donna turned wondering eyes upon him, he shrugged. "See," he said, "my mother's people used to fish the traditional ways, with net and line. Then war came. That's when she met my father. My people learned to fish by dropping dynamite on the reefs. It made fishing a lot easier and nobody figured what the end result would be, even when there were fewer and fewer fish. I used to hear the men talking about it.

"Then my mother died and I lived with my grandparents. But after the war, my father came back and took me to Scotland to go to school. My grandparents were glad to let me go because by that time the reefs were gone, and all the fish with them.

"My mother's people were already hungry even then. And now there are twice as many people as there were when I was a kid."

"Is there a way to bring the fish back?"

"No. Well, I don't know. You know how long it takes to build a reef. It'd be simpler to stop having so many people—and that's something we could do right away." He shrugged. "But *that's* something people have to decide for themselves." Then he grinned.

"But maybe I can become a professor, like Dave, and go back there to help some day."

Approaching the deep trench, Frill, of all creatures, raised a major problem. At one of the rest stops, she announced, "This is what I wish to do," and projected an image of herself in the shuttle.

"Now, wait a minute!" Dave was aghast. "Frill," he marshalled his arguments. "First of all, we can't put a tank in the shuttle . . . ."

"You would not need to," she was perfectly tranquil. "I can get into one of those things humans wear in cold waters."

"A wetsuit?" Kate tried to sound as doubtful as possible. But her heart wasn't in it, and both Frill and Fan immediately knew it.

Now Fan raised a protest. "It is far too dangerous. The pressure at such depths will kill. Besides, we cannot stay out of water for any length of time. The risk is too great. In any case," in a total turnabout, "I should be the one to go."

Unexpectedly, Dr. Vee entered the argument. "That is completely illogical. First, the shuttle allows descent to great depths—that is its purpose. Second, use of the wetsuit permits encasement in water. And third, you are far too large to fit in the shuttle."

At this point, Fan went into an aquabatic display of temper, the first that the humans had ever seen in a cetasapien. The water fairly boiled around the *Turtle*, the other CTs scattering to a safe distance. It was obviously not a good idea to get in the way of an angry, frustrated cetasapien male.

But Dr. Vee and Frill remained by the bubble.

"Isn't that so, Professor?" Dr. Vee said imperturbably.

"Eh?" Dave was at first distracted. "Oh. Yes, well . . ." Both attracted to the idea of having Frill as a partner in the depths, and concerned for her welfare, Dave felt impelled to point out the difficulties.

"Frill, being encased, confined in that way, is so totally alien to your experience. At best, it would be extremely uncomfortable, at worst, disorienting."

"I will not become disoriented," Frill was totally confident.

"You know," Dave made another try, "you have lived a very open, free life. Confinement of any kind, especially in such close quarters, combined with physical discomfort, may be more than you can imagine. Stress is something with which your kind is not familiar."

A snort from Dr. Vee. "What do you suppose we encounter every day in the healing lagoon, young man?" Dave noted he had suddenly been demoted.

"Stress, even when there is no physical damage, is precisely one of the problems we have to deal with—and increasingly, with the encroachment of humans. So, if you are prepared to take the risk," grandly, "certainly we are."

"I am not." Fan had returned.

Frill turned on him. "This is a priceless opportunity to study our world," she said severely.

Kate decided this was an emotional issue in which the humans had best not intervene. She was reasonably certain Frill would get her way. As captain, she, of course, could not have dived in the shuttle herself. Frill, Kate judged, was a good second bet. So she simply said, "It's up to all of you. If the decision is that Frill should go, we will find a way."

However, getting Frill into the shuttle proved to be a real drama. She was perfectly capable of swimming up into the lock by herself, but it was not designed to be opened at 8,000 feet. Kate elected to get Frill in at a shallow depth so that she could be assisted by one of the aquanauts. Two more—Teiji and Paul—waited in the dive room to help her out of the lock chamber and into a very baggy "wetsuit" which the crew had been carefully constructing for days.

Next, there was the problem of getting Frill's 250-pound bulk up the ladder and into the shuttle, where Dave waited, already ensconsed so that he could help get Frill settled.

Teiji had rigged a winch to haul the cetasapien up: the wetsuit itself, although well sealed wherever it met her body, was actually more like a water cradle, adding its own very necessary weight. Dave had extra water bottles handy to spray her protruding head, flippers and flukes.

Although the shuttle had a 3-day capacity, the dive was scheduled to last only 12 hours. Even Frill judged she could not remain encased longer than that. So, to save time, while she was being eased into the shuttle, the *Turtle* descended to 12,000 feet preparatory to the launch, leaving Fan prowling restlessly at 8,000 feet, only partially reassured by the continual flow of messages from Frill to her anxious mate.

When everything was secure, Teiji took his position at the controls of the robot probe, the "Bug," which would accompany the shuttle. Both the shuttle and the Bug would be sending back images for as long as possible but the Bug was mainly there to focus on the shuttle itself.

Frill's personal images to Fan faded at around 22,000 feet.

The *Turtle* was instantly flooded with the anxiety radiating from Fan like a smothering blanket and rapidly spreading to the other cetasapiens. Nor were the humans immune from such distress.

"Paul!" Kate snapped.

"Right." Paul found himself urgently reassuring the big male, sending long, steady signals that would reach through the 4,000 feet that separated them. Fan calmed down somewhat. The truth was, and he knew it, that at a depth of 36,000 feet as little as possible was being left to chance. The humans had told him that this ocean trench was deeper than their highest mountain was high. So they were being very careful.

Although oceanographers had known for some time that life existed even at the extreme pressures encountered at these depths, the whales had never descended here so nothing was known to the cetasapiens of this part of their oceans. The giant squid, who lived at depths greater than any other complex creature, had few interests beyond their own immediate needs, and reported in only the vaguest of terms.

Frill meanwhile was tremendously excited, even exalted. Fan had caught some of that feeling before she faded out. It was a heady thought, to be the first cetasapien ever to see the weird life-forms the lights of the two vehicles revealed.

Frill forgot, for a time, the discomfort of her position. Dave kept up a running commentary, some of which reached the *Turtle* in garbled form, but all of which was recorded in coordination with the video they were shooting. And although it was pitch black, the shuttle's sensors were recording the terrain and preventing them from crashing into the mountains.

After videotaping all the life-forms they could find, Dave elected to do a slow motion exploration of the Trench, at least for as much time as they had. He wanted to leave a margin of safety for the return trip despite Frill's protests that she was perfectly all right. If it had been up to her, they would have stayed the full three days!

There was not much food in the icy waters. The elastic stomach allowed a meal, swallowed by a gigantic mouth, to last for several days. Many creatures developed bioluminescent organs, creating their own dim and sometimes flashing light, to attract prey, or perhaps to frighten other predators.

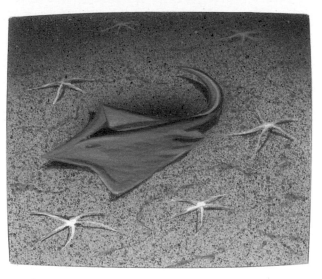

### Cirrate Octopus

Dave's commentary answered a number of Frill's eager questions. He pointed out that abyssal creatures generally adapted to the extreme pressures of the deeps by developing enormous eyes—or no eyes at all, since nothing could be seen in the stygian darkness— and also, many had expandable stomachs.

### Deep Sea Skate

From the bottom, below 36,000 feet, the Bug scooped up mud and silt, storing the glop in various compartments. Later Dave would examine it all under a microscope to search for the minute life-forms he was confident they would find.

### Medusa Jellyfish

### Glass Sponge

### Sea Cucumber

Oddly enough, the jellyfish species, seemingly so fragile, did well in the deeps, perhaps because they embraced the problem of pressure, their bodies being formed largely of water. Still others, like the skates, buried themselves in the bottom detritus.

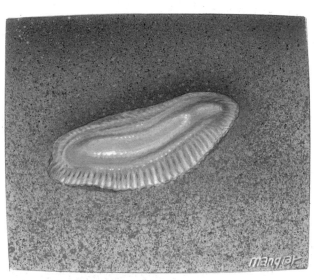

Dave stayed as long as he could but time eventually ran out and they began their ascent. All was going well. They had reached 13,000 feet, within 1,000 feet of the *Turtle*, when they were buffeted by violently moving water, at that depth an alarm signal in itself that something was seriously wrong.

In moments they were battered by slowly tumbling stones hitting the exterior of the shuttle.

"We're right on the edge of a landslide," Dave said calmly. But reverberations numbed their senses as the shuttle lurched, coming to an abrupt halt. Although strapped in and cushioned, Dave was knocked unconscious when the transmitting unit jolted loose and landed on his head. Frill, buffered in her watery bed, was better protected, and fortunately nothing pierced her couch.

But all electronic communication with the *Turtle* had ceased.

Then a small sound penetrated the deathly silence as the Bug floated by, fixing their position in its lens.

In the *Turtle*, the crew was dead alert and very tense. The image from the Bug showed the shuttle caught in rocks. She wasn't held by much, but could not move. And the break in communication was terrifying.

Then Paul picked up from Fan. "Oh, no!" he muttered. And loudly, "Fan! Stay where you are! You can't do any good by coming down. The pressure here will kill you for sure!"

"He's at 9,000 feet and coming," from Donna.

"Stop him, Paul," Kate said urgently. "Say if he wants to help to stay put and try to contact Frill."

Paul sent a loud burst of rackety sound, like a slap in the face, to stop the cetasapien's headlong descent.

"Okay, right. Stay put, Fan. *Contact Frill. Contact Frill.*" Pause. "He's in touch: she's okay but Dave is hurt—unconscious." Pause. "Frill knows where the center of the stuff holding them is located. She can sense it."

"Teiji? Could the probe do something? Could Frill help in any way?"

Teiji said slowly, "She'd have to transmit through Fan, Captain. Fan could tell us. And it would have to be very, very precise to avoid another landslide that might bury them."

"It's the only chance we have. Do it, Teiji."

Working in unison, Frill, Fan, Paul and Teiji began the critical task of removing one rock at a time from the rubble covering a portion of the shuttle.

Jay was at the pilot station, with Kate watching each maneuver like a hawk. Donna kept track of Fan's position, conscious of the stress he was feeling, sending, as best she could, reassurance both to him and to the cetasapiens above.

Teiji's work was the most meticulous: he could not afford to make even a fractional error, but Frill's senses, made all the more acute by their danger, proved to be minutely accurate. With infinite delicacy the smallest fragments of rubble were finally removed.

Kate, in the meantime, had been wondering how to give Frill directions to fly the shuttle, especially in its damaged condition. The cetasapien simply did not have the physical equipment. She was, however, able to squeeze the water-bottles brought for her benefit, and for Frill, in any emergency, but especially where health was at risk, water was primary. Determinedly she somehow twisted around and sprayed David's head.

He came to, groggily muttering, "My God, we've sprung a leak," and on the instant

realized they could not possibly be alive if that had happened. A few minutes, and he announced himself able to pilot.

They still did not have communication, but the crew back at the *Turtle* already knew he was conscious through the Frill-Fan link, and shortly they watched as the shuttle slowly and shakily made its way back up to dock with the mothership.

With the shuttle safely locked in, Dave was immediately escorted to sickbay, protesting all the way. Paul and Teiji and Jay managed to get Frill down and into the tank in the lock chamber. There she had to wait while they climbed back up to where she at last, and very thankfully, could be released to join Fan and her impatient family.

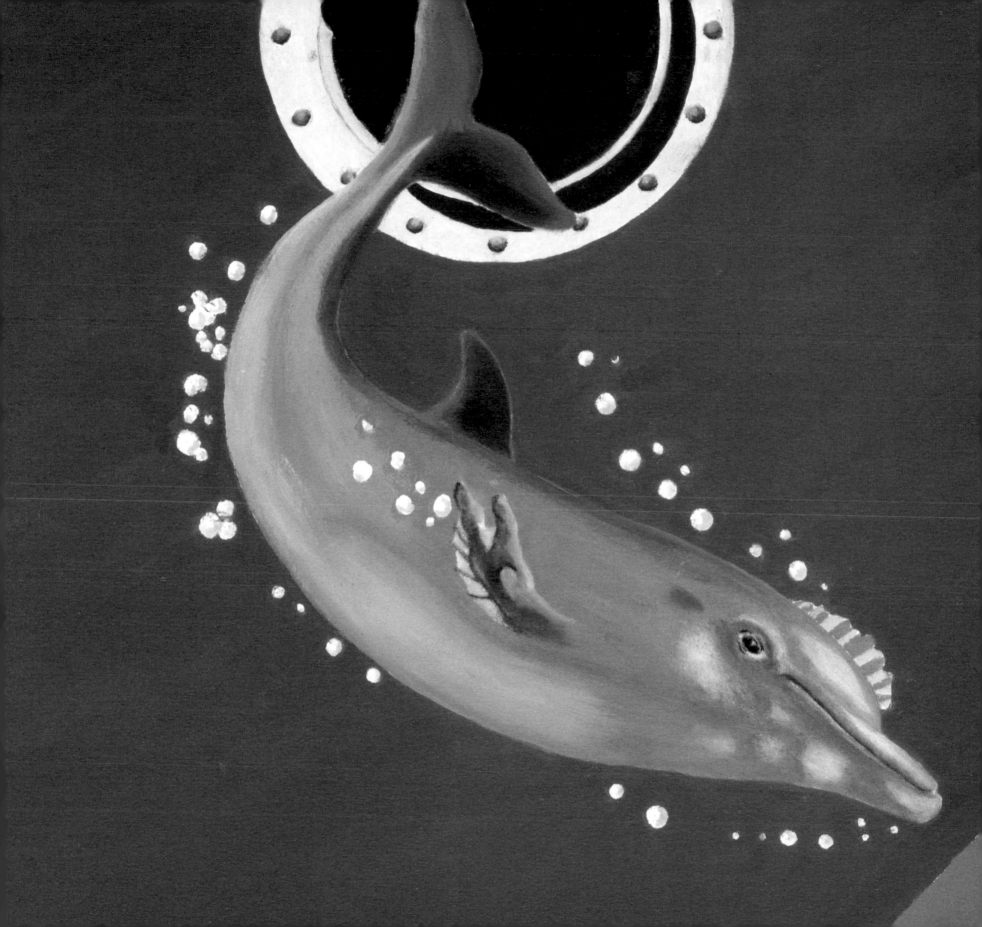

The cetasapiens were very excited by all that had happened, and needed to work off their delight in Frill's safety
and their pride in Fan's bravery: an ebullient mini-ballet
took place as they spiralled up toward the surface.

Among the humans too, there was intense relief.
Dave, it turned out, was not seriously injured, so
amid general rejoicing the whole expedition now
set off for homebase, where anxious families had
long been waiting to welcome them.

The returning *Turtle* with its full complement of cetasapiens received a tumultuous welcome. A stream of greeters poured out of the great cave entrance through which the *Turtle* had first entered homebase.

A chorus of happy sound—hoots, whistles, wails, chortles, clicks and tappings—exploded within the command module as the sea creatures welcomed one another and the *Turtle*.

This was a very different arrival than their first encounter had been. For all of them, this was truly coming home.

Once the *Turtle* was secure in the lagoon, Donna rushed off to find her baby CT, only to be greeted by a large, well-muscled fellow, more than capable of whirling her giddily around in the water—which, to his great delight, he proceeded to do. Obviously the cetasapiens, like their surface brethren, grew and gained weight with enormous rapidity. But, Donna thought wistfully, he'll stay by his mother's side for at least another year or two. Instantly sensing her longing, Aunt Em came over to give her a watery hug and fuss over her.

Meanwhile, Teiji and Paul had joined Fan, their shared experience in rescuing Frill and Dave having created a strong bond. They were happily ensconsed in Frill's homecave. They would have welcomed the presence of the others, but Jay, Dave and Kate were in the healing lagoon, along with Dr. Vee, enthralled to have his precious medicinal

sponges installed in their future home, there to flourish and thrive—a gift the humans had brought to the cetasapiens.

Everyone felt the need for a quiet time, time to absorb all they had seen, time to recover from the tensions and excitements of their recent experience in the Marianas. So for a few days, the aquanauts simply relaxed among the cetasapien families.

Paul was delighted to find that his hanging xylophone had been imitated—sort of. Some enterprising CTs had collected bits of metal pipe that now hung at entrances to homecaves. Typically, each sound identified a different family, and each plangent note harmonized with all the others. The younger CTs had a great game, racing around to see who could clang all the "bells" fastest.

One evening Dave said quietly to Kate, "You know, the implications of all that has been happening to us are just beginning to really hit home."

"I know. I've been wondering for some time what our friends expect us to do. We've all been so engaged with the day-to-day drama that the future has seemed too remote to bother with."

"The future is why we were brought here." Paul had joined them.

Dave felt a glow of pride. "You're right, of course. And especially so for you and Donna."

"Well, I don't know. I think they have longer range objectives than just the lifetimes of a pair of humans. The entire way of life in the system of oceans—one ocean really—and their method of communication, also open and total—not one word at a time, like ours—has to have a built-in sense of balance. You might say they are in a constant state of adjustment in order to maintain that balance. That's why human excesses puzzle them so much. They *know* we are intelligent. They can't figure out why we behave so stupidly."

"Again you're right. D'you know what Dr. Vee said to me the other day? We were talking about food sources—theirs are getting as scarce as ours—and he remarked that they have always killed only enough to eat, and never their own kind. 'You humans,' he said, 'you kill other humans by the million and you do not even eat them. Do you over-produce simply in order to have something to kill?' It shook me, I can tell you."

Paul sighed. "Good question. From his point of view he's just trying to find a logical answer to two inexplicable forms of behavior. The problem is that just doing our thing, the way we live, is a real threat to them. And to us."

"We'd better find out what they have in mind." Kate, always practical. Dave agreed and together they called a meeting to see what the next step was to be.

A full gathering
of the cetasapiens who
lived in the lagoon and all the
transient guests who visited for
one reason or another was awesome.

David got things going. "My friends,
the last time we all met this way, we were
strangers, seeking the reason why we had
been brought here. And we agreed to make a
long journey with you. Now we must make
another journey, back to our own kind. And
again we need your guidance. For we must carry
word back to them of what we have learned
here. We seek your advice on what we should
tell the upper world."

Frill called on various individuals for opin-
ions. In the relaxed fashion of the oceans, the
meeting took on the air of an easy discussion
between friends, even though remarks came one
at a time, in bursts, and with only generalized con-
nections. But that too, was characteristic of the
oceans. The humans listened.

There must be no deception. Humans are very complicated creatures, very contradictory, easily confused. Therefore it is necessary that we be direct and simple.

We must assume that humans value life, although there is much evidence to the contrary. If they do, then they should learn to value balance, for without it life cannot exist.

Life is joy. Life is balance. Life is diversity.
The ocean is the wellspring of life.
The coastlines are the bridge to life.

All life, on Earth, and in the universes beyond, is unique, singular, and all life-forces balance each other. Whales are the critical sensory points for this planet, for we are the global communication system. We have the brain capacity to absorb, retain and broadcast ever-changing knowledge, and our strength and senses allow us to transmit life-sustaining information over long distances.

A myriad species have lived in the oceans for millennia and have never experienced population problems. Nature maintains balance. This could be true for all air breathers if humans would allow it. But balance on land cannot occur without the oceans, for oceans and land are themselves in balance.

For any to survive, all must survive. Perhaps because humans have such short lifespans they cannot think beyond a generation or two?

Quietly Frill took over: "The whale communication system, for humans, and for all life, is the vital element in holding balance for land- and sea-dwellers. So. Tell your people their experiment is a triumph. Humans and cetasapiens have established communication. Tell them who inhabits the oceans. Do not tell them where. Not yet. You do not, in fact, really know, so there is no deception. There is balance in all things."

She paused, but presently continued. "Nature offers so much: change, and the ability to adapt. Life and living are a constant process of adaptation, a joyous balancing act which demands our best. That is why our loudest singers are also our most potent males, for they sing of life, they sing the Song of Earth." Gently she added, "It is your song fully as much as it is ours. But only you can make it so."

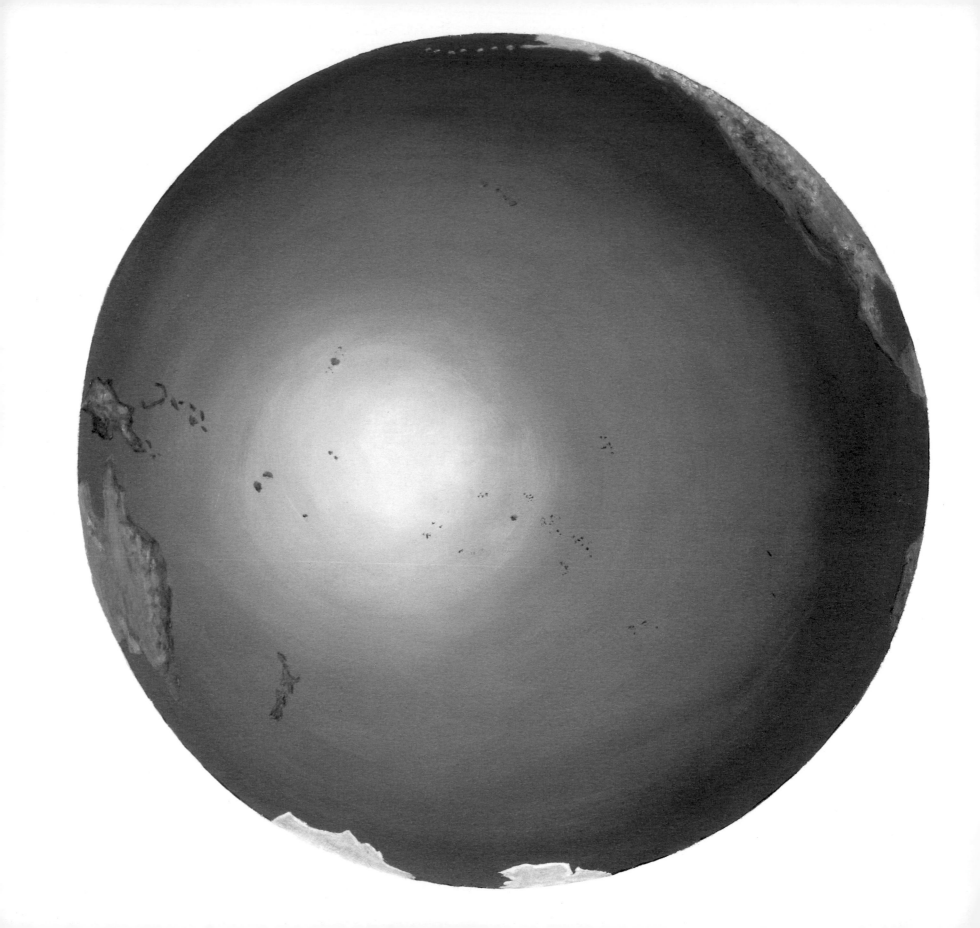

The gathering dispersed, slipping gently back into the waters of the lagoon. Frill and her family remained, and the humans continued to sit quietly on the *Turtle*, reluctant to go back inside. Suddenly Kate realized why. All of them were now confronted with the fact of separation. Kate had spent a lot of her life in partings. Her throat ached for herself and her friends at the thought of yet another break with what had become dear and familiar. She wondered how Frill and her family might be feeling.

"Frill? Could we come over to the homecave? Tomorrow we'll need to spend most of the day preparing for departure. I think we'd all, well, just like to be—with you, for a while."

The flood of affection that enfolded them was almost physical. Donna gulped and rushed into the *Turtle*. The routine of getting into dive gear helped them all to safer emotional ground. As each of them entered the water Frill's family crowded around, gently jostling and nuzzling, and eventually leading them off through the well-known tunnel leading to the homecave.

Arrived at the cave, the aquanauts shucked their tanks and made themselves comfortable. Dave judged it best to get to practical matters before everyone was overcome by the sense of imminent departure hanging over them all.

"That was a remarkable meeting," he said easily. "And I've been trying to think how we could go about conveying the importance of what we've learned to the upper world."

"You have brilliant technology," observed Dr. Vee. "It should be simple. You can put the whole story on a screen to show to millions, can you not?"

"That's right," Frill agreed. "You must function with the tools of your element and your culture. Humans have different strengths from ours. Use your strengths. Use your resilience. Use your technology. And remind all to use it for the right purposes, yes?"

She too seemed to think it would be easy. Jay voiced discouragement. "I dunno. Humans are stubborn, rigid. They grow up with one set of ideas and it's hard to shake them loose."

Aunt Em said fiercely—or at least what passed as fiercely in a cetasapien, "Talk to the young. It is their lifespans that are threatened. Even the elders, once they really understand this, would be willing to listen." She slid over and placed her body protectively in support of Donna, who turned at once and buried her face in the soft folds under Aunt Em's head. Teiji's broad shoulder pressed up against Fan.

Dr. Vee turned sadly to Jay. "I shall miss you, my friend," he said. Dave suddenly found himself very glad that he was holding Kate's hand in a firm grip.

And Paul. Very gently Paul tapped the heavy copper tube that hung at the entrance, half in and half out of the water—like our lives, he thought. He tapped again. Soft harmonics from other caves joined his. Presently, to his astonishment, a melody began to emerge—the same plaintive seasong that he had played on the night before their departure on this adventure. The cetasapiens had listened well.

Paul moved over to Donna. "Come on, Donna," he said gently, and helped her into her dive gear. The others stirred themselves, and all the cetasapiens slipped into the water to accompany them back through the echoing tunnels to the *Turtle*. As they entered the ship, the voices of the whales reverberated through the caverns, singing the home song for their human friends.

Much later, with Donna at last in an exhausted sleep, Kate stood at their cabin viewport, gazing into a world she had once distrusted, and now passionately loved. Frill was right there, gazing back. Through the shadows of grief, Kate felt a well of happiness and reassurance.

Fists clenched up against the window, Kate said aloud, "We'll be back! Oh Frill, we *will* be back!"

## ACKNOWLEDGMENTS

Each of the artists who contributed to this book had something special to offer, and gave unstintingly of both effort and talent: we are grateful most of all for their supportive enthusiasm and willingness to cooperate with one another and with the author in the creation of this new form of graphic reading experience.

**Lloyd Birmingham** is an oceanographer who vividly charts the oceans, and has worked extensively for the *National Geographic*. Pps. 50-51, 86-87, 102, 103.

**Steve Brennan** art majored in 1987 and has already produced over 250 works for book covers, postage stamps, magazines and television. Pps. 127, 128, 129, 130, 131.

**Joseph DeVito**, known for his dramatic *Doc Savage* covers and for breathtaking compressed action, is also an accomplished sculptor. Pps. 82-83, 84-85, 120-121, 122-123, 124-125.

**Dave Henderson**, an art major from Boston University, is an expert technical artist who enjoys landscape painting and is also a skilled craftsman and woodworker. Pps. 48-49, 94, 95, 104-105.

**Carol Inouye**, formerly art director of Ballantine Books, now lives in Maine, running her own design studio, creating art books and illustrations for children's books and visual works such as *The Native Americans*. Pps. 68-69, 70-71, 72, 73, 74, 75, 76-77, 78-79, 88-89, 90-91.

**Robert Larkin** prefers to remain a master of all styles, and works in a wide range of media. Pps. 126, 140-141, 142, 144-145, 146-147, 148-149, 150-151, 152-153, 156-157.

**Gilles Malkine**, landscape artist who storyboarded *The Secret Oceans* and contributed several original subjects: he is also an architectural artist and designed the *Turtle*, stem to stern. Pps. 7, 38-39, 42L, 92-93, 106-107, 142R, 143, 154, 158, 159.

**Jeff Mangiat**, in demand even before leaving art school, created the cover painting, and is expert in engineering, space, and sport art. Pps. 28-29, 96-97, 98-99, 100-101, 132-133, 134-135, 136-137, 138, 139.

**Thomas McNeely** is a Canadian who excels in paintings of the Arctic world, wild life, Canadian Indians, and character studies. Pps. 1, 2, 10-11, 43, 45R, 54-55, 56-57, 58-59, 60-61, 62-63, 64, 65, 66-67, 80-81.

**Davis Meltzer** comes from a family of fine artists and is known for his NASA posters and *National Geographic* art, specializing in accurate portrayal of the natural world. Pps. 52, 53, 108-109, 110-111, 112-113, 114, 115, 116-117-118, 119.

**Charles Passarelli**, mechanical expert (he collects old cars), specializes in dramatic human action and in luminous land-and waterscapes. Pps. 8-9, 12-13, 14-15, 16-17, 18, 19, 20-21, 22-23, 24-25, 26-27, 155.

**Jeffrey Terreson**, specialist in imaginative land-and seascapes, and people in action. Pps. 30-31, 32-33, 34-35, 36, 37, 40-41, 42R, 44, 45L, 46-47.

# BIBLIOGRAPHY

*Bantam World Atlas, The.* New York, NY: Bantam Books, 1989.

Blake Publishing. *The Nature Series*; and *The Habitat Series.* San Luis Obispo, CA: Blake Publishing, 1989.

Borgese, Elisabeth M. *The Drama of the Oceans.* New York, NY: Harry N. Abrams, 1975.

Brower, Kenneth. *Realms of the Sea.* Washington, DC: National Geographic Society, 1991.

Cochrane Amanda, and Karena Callen. *Dolphins and Their Power to Heal.* London, England: Bloomsbury Publishing, Ltd., 1992.

Conley, Andrea. *Window on the Deep.* New York, NY: Franklin Watts, 1991.

Cousteau, Jacques. *The Ocean World.* New York, NY: Harry N. Abrams, 1989.

Cousteau, Jacques. *The Ocean World of Jacques Cousteau: Oasis in Space* (Vol. 1). Danbury, CT: Danbury Press, 1973.

Cousteau, Jacques, and Philippe Diole. *The Whale: Mighty Monarch of the Sea.* New York, NY: Doubleday & Co., 1972.

Cromie, William J., and William H. Amos. *Secrets of the Seas.* Pleasantville, NY: Reader's Digest Association, 1972.

Daniels, George, ed. *Volcano: Planet Earth.* Alexandria, VA: Time-Life Books, 1982.

Doubilet, David. *Light in the Sea.* Charlottesville, VA: Thomasson-Grant, 1989.

Doubilet, David. *Pacific: An Undersea Journey.* Boston, MA: Bulfinch Press, 1992.

Ganeri, Anita. *The Oceans Atlas.* London, England: Dorling Kindersley, 1994.

Guerrini, Francesco. *The Great Book of the Sea: A Complete Guide to Marine Life.* Philadelphia, PA: Running Press Books, 1993.

Hamilton-Paterson, James. *The Great Deep: The Sea and its Thresholds.* New York, NY: Random House, 1992.

May, John, ed. *The Greenpeace Book of Dolphins.* New York, NY: Sterling Publishing Co., Inc., 1990.

*National Geographic.* Oceanographic articles. Washington, DC: National Geographic Society.

Newbert, Christopher. *Within a Rainbowed Sea.* Hillsboro, OR: Beyond Words Publishing, 1990.

Scheffel, Richard L., ed. *Nature in America.* Pleasantville, NY: Reader's Digest Association, 1991.

Snyderman, Marty. *Ocean Life.* Lincolnwood, IL: Publications International, Ltd., 1991.

Steel, Rodney. *The Concise Illustrated Book of Sharks.* London, England: Grange Books, 1993.

Steene, Roger. *Coral Reefs: Nature's Richest Realm.* New York, NY: Bantam Doubleday Dell, 1990.

Stevenson, Robert E., and Frank H. Talbot, eds. *Oceans: The Illustrated History of the Earth.* Emmaus, PA: Rodale Press, 1993.

Thorne-Miller, Boyce. *Ocean.* San Francisco, CA: HarperCollins San Francisco Group, 1993.

Throckmorton, Peter, ed. *The Sea Remembers: Shipwrecks and Archaeology.* New York, NY: Weidenfeld & Nicolson, 1987.

Wells, Sue, and Nick Hanna. *The Greenpeace Book of Coral Reefs.* New York, NY: Sterling Pub. Co., 1992.

Wilkinson, Peter, ed. *Wildlife Photographer of the Year: Portfolio Two.* London, England: Fountain Press, 1993.

Williams, Heathcote. *Whale Nation.* London, England: Jonathan Cape Ltd., 1991.

*World Atlas.* Maplewood, NJ: Hammond Inc., 1976.

Wu, Norbert. *Beneath the Waves.* San Francisco, CA: Chronicle Books, 1992.

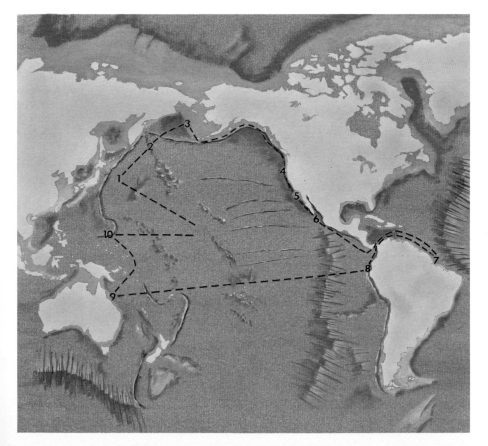

## Map of the Route of the *Turtle*

| 1 | Ghost Nets | 6 | Baja |
|---|---|---|---|
| 2 | Emperor Mounts | 7 | Amazon Delta |
| 3 | Arctic | 8 | Sunken Ship |
| 4 | Oregon Coast | 9 | Great Barrier Reef |
| 5 | Kelp Forests | 10 | Marianas Trench |

'till we meet again